D1415757

LAB MANUAL TO ACCOMPANY BOCTOR'S ELECTRIC CIRCUIT ANALYSIS

MOHAMED GHORAB

SHIVA PRABHU

Ryerson Polytechnical Institute

PRENTICE-HALL, INC., Englewood Cliffs, New Jersey 07632

Editorial/production supervision: Colleen Brosnan
Manufacturing buyer: Carol Bystrom
Page layout: Fran Kasturas

© 1987 by Prentice-Hall, Inc.
A Division of Simon & Schuster
Englewood Cliffs, New Jersey 07632

All rights reserved. No part of this book may be
reproduced, in any form or by any means,
without permission in writing from the publisher.

Printed in the United States of America

10 9 8 7 6 5 4 3 2

ISBN 0-13-247420-4

Prentice-Hall International (UK) Limited, *London*
Prentice-Hall of Australia Pty. Limited, *Sydney*
Prentice-Hall Canada Inc., *Toronto*
Prentice-Hall Hispanoamericana, S.A., *Mexico*
Prentice-Hall of India Private Limited, *New Delhi*
Prentice-Hall of Japan, Inc., *Tokyo*
Prentice-Hall of Southeast Asia Pte. Ltd., *Singapore*
Editora Prentice-Hall do Brasil, Ltd., *Rio de Janeiro*

CONTENTS

© 1987 by Prentice-Hall, Inc., A Division of Simon & Schuster, Englewood Cliffs, N.J. 07632. All rights reserved. Printed in the United States of America.

PREFACE

The important objective of any laboratory session is to reinforce the understanding and the use of theoretical background through first hand experience with relevant experiments and calculations. This laboratory text is written with the primary goal of addressing the above objective and as well to provide the student with the skills needed in using the equipment properly.

The experiments covered in the text are designed to complement the text "Electric Circuit Analysis' by Dr. S.A. Boctor. They have been developed and tested over the past few years at the authors' institute and proven to be effective in aiding the students in better understanding of the basic concepts in Electric Circuit theory.

Each of the above experiments begins with a reference to a specific chapter and section in the text book, followed by a pre-lab assignment related to the objectives of the experiment and is meant to enhance the students' appreciation of theory versus practice. The experimental procedure is written in "steps-format" accompanied by appropriate tables and graphs. This format allows the student more time to examine the experimental data and analyze the results. Each experiment concludes with a set of questions designed to achieve a better understanding of the concepts examined.

This laboratory text is readily adaptable to a two-semester course in Electric Circuit Analysis.

We are grateful to our colleagues at our institute for their suggestions and help in improving the laboratory text over the years, especially to Professor John Van Arragon, whose work has provided the necessary basis for many of these experiments.

INTRODUCTION TO ELECTRICAL MEASUREMENT

0.1 Definition of Terms:

In making measurements (voltage, current, resistance, etc.), we employ a number of terms which need to be accurately defined.

These terms are:

a) Sensitivity: "a measure of the smallest quantity (voltage, current, etc.) that will produce an observable deflection" (using deflection-type instrument).

b) Accuracy: "the degree to which the measured value approaches the true value", usually expressed as a percentage of full-scale deflection (fsd). For example, a voltmeter with a fsd of 100V may have its accuracy stated as ± 2%. When the pointer of this voltmeter indicates 50V, the actual measured voltage must be taken as 50V ± 2% of fsd, i.e. the measured voltage is somewhere between 48V and 52V.

c) Precision: "the degree to which a measurement is sharply defined", expressed by the number of significant figures in the reading. It is also defined as the smallest measurable difference between two readings.

d) Linearity: "the degree to which a graph of the actual value of measured quantity, versus the reading of the instrument, deviates from an ideal straight line drawn between the end points of the actual curve", as shown in Fig. 0.1

© 1987 by Prentice-Hall, Inc., A Division of Simon & Schuster, Englewood Cliffs, N.J. 07632. All rights reserved. Printed in the United States of America.

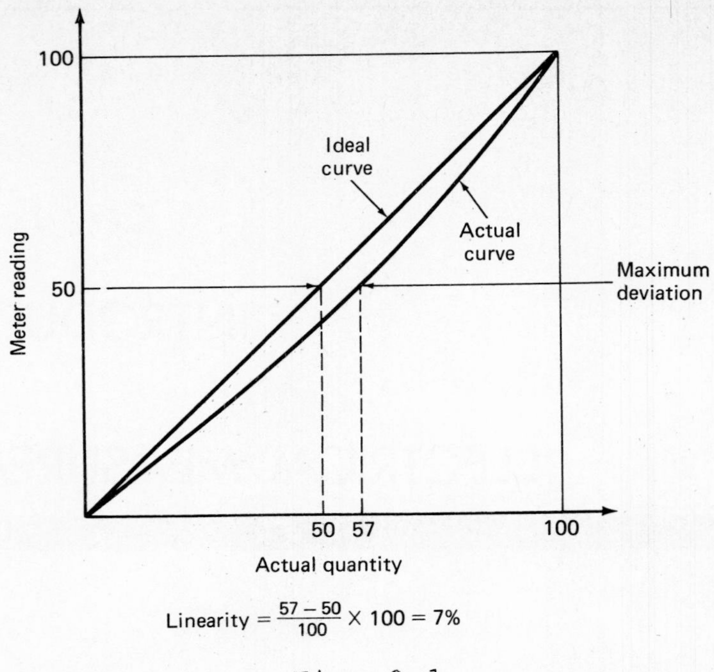

Fig. 0.1

0.2 Measurement Pitfalls:

Here are a number of tips for interpreting measured data:-

a) Meter accuracy is usually specified by the allowable error as a percentage of fsd. Thus, a reading of 15V, on the 100V ± 2% scale, must be interpreted as having the limits of 15 ± 2V or 15V ± 13.3%. Obviously, for the greatest measurement accuracy, the best range to use is the one that gives the largest deflection, not exceeding the fsd (full-scale deflection).

b) Meter calibration often changes abruptly with range changes, without any actual change in the measured quantity taking place. Therefore, whenever a string of readings is to be taken, it may be wise to take all readings using the same range.

c) Beware of subtracting two almost equal pieces of data. The uncertainty associated with each reading can make the result of subtraction almost meaningless. For example, 40V and 45V readings on a 100 ± 2% scale could be either (47-38) = 9V or (43-42) = 1V; the ideal result, however, is 5V. It can be easily shown that the limits of the error here are ± 80%.

0.3 Reading of Deflection-Type Meter:

$$\text{The meter reading} = \text{Range} \times \frac{\text{Deflection}}{\text{fsd}}$$

Example 0.1

Determine the reading of the multimeter, shown in Fig. 0.2, for each position of the range selector.

Range Selector Position:

a - 10V
b - 250V
c - 100 μ A
d - 50 mA

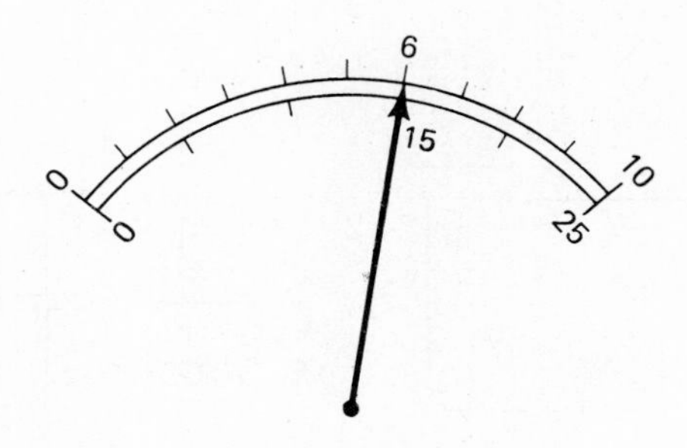

Fig. 0.2

0.4 Current Measurement:

The instrument used to measure electric current is called an "ammeter." Ignoring its internal components, a deflection-type DC ammeter consists basically of: a calibrated (linear) scale over which a pointer is deflected to indicate the measured current, two terminals identified by a + and -, and a range switch to select the current range of each particular measurement.

An ammeter must be connected so that the current to be measured flows through the meter. Consequently, AMMETERS ARE ALWAYS CONNECTED IN SERIES WITH THE COMPONENT IN WHICH THE CURRENT IS TO BE MEASURED, as shown in Figure 0.3.

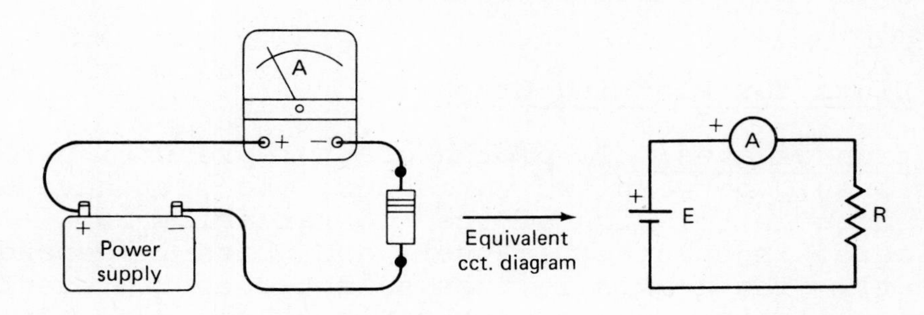

Fig 0.3

© 1987 by Prentice-Hall, Inc., A Division of Simon & Schuster, Englewood Cliffs, N.J. 07632. All rights reserved. Printed in the United States of America.

0.5 Voltage Measurement:

The instrument used to measure the voltage difference between two points in an electric circuit is called a "voltmeter." The basic deflection-type DC voltmeter is similar in appearance to an ammeter, i.e., the instrument has a + and - terminal, a range switch, and a pointer which moves over a linearly-calibrated scale in volts.

VOLTMETERS ARE ALWAYS CONNECTED IN PARALLEL WITH THE CIRCUIT-NODES ACROSS WHICH THE VOLTAGE DIFFERENCE IS TO BE MEASURED, as shown in Figure 0.4.

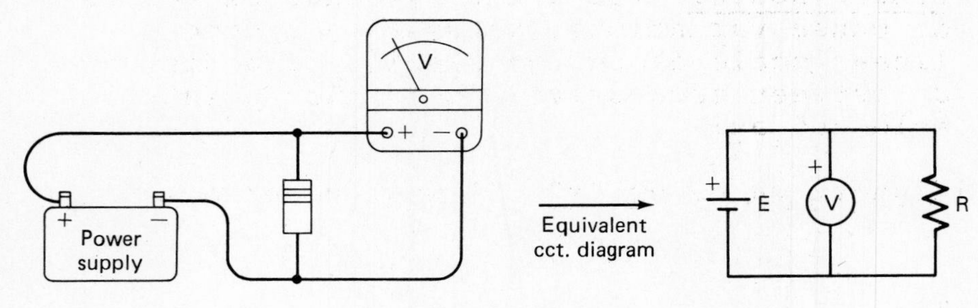

Fig. 0.4

0.6 Resistance Measurement:

Resistance is measured by means of an "ohmmeter." The Ohmmeter has its own internal DC source for the purpose of passing a current through the resistance to be measured. The resulting current (which is inversely proportional to the resistance) causes the pointer to move over a nonlinearly-calibrated scale in ohms. Notice that the ohmmeter scale is quite different from that on the other meters: it is nonlinear; zero deflection corresponds to infinite ohms, and full-scale deflection (fsd) corresponds to zero ohms.

AN OHMMETER MUST NEVER BE CONNECTED TO ANY PART OF AN EXTERNALLY-ENERGIZED CIRCUIT.

The Ohmmeter measures resistance most accurately when its pointer position is around the mid-scale range.

0.7 Hints for Plotting Graphs:

a) Selection of axes: To plot a graph representing the relationship between two variables, one must have a clear idea as to which of the two variables is considered independent (cause), and which is dependent (effect). The x-axis is then selected to represent the cause, while the y-axis represents the effect. In current (I) versus voltage (V) graph, current is dependent on the magnitude of the voltage, i.e. voltage is the cause for the current (effect) to flow.

b) <u>Selection of Observation points:</u> With unlimited time and energy at hand, one might consider taking many observations (measurements) to plot a graph; the more observations taken, the more accurate is the resulting graph. However, for a limited number of observations, $\underline{n}$, we wish to select the observation values along the cause-axis (x-axis) at which we take our measurements of the effect (y-axis), and which also provides optimum graph accuracy. Our choice will depend on the width of the range of values of "x" over which we wish to measure "y."

(1) <u>Linear-scale:</u> for a narrow range of values of cause variations, we usually choose a linear scale along the x-axis. The distance, $\underline{d}$, between successive observation points is selected as:

d = [highest value of "x" - lowest value of "x"] / (n-1)

Example 0.2

Obtain the observation values along the x-axis that provide optimum graph accuracy. Given:

$0 \leq x \leq 10$, and n = 6 points

<u>Solution:</u>

$d = (10-0)/(6-1) = 2$

Thus, the values of "x" are: 0, 2, 4, 6, 8, & 10.

(2) <u>Logarithmic (non-linear)-scale:</u> For a wide range of values of cause variation, we usually choose a logarithmic scale along the x-axis.

The ratio of two successive observation points, $\underline{r}$, is selected as:

$\underline{r}$ = [highest value of "x"/lowest value of "x"]$^{\frac{1}{n-1}}$

Example 0.3

Obtain the observation values along the x-axis that provide optimum graph accuracy. Given:

$10 \leq x \leq 4000$, and n = 9 points

© 1987 by Prentice-Hall, Inc., A Division of Simon & Schuster, Englewood Cliffs, N.J. 07632. All rights reserved. Printed in the United States of America.

Solution:

$$r = [4000/10]^{(1/8)} = 2.11$$

Thus, the values of "x" are 10, 21, 45, 96, 200, 423, 894, 1891, & 4000.

0.8 Errors:

It is not possible to measure any quantity with perfect accuracy and as such, it is important to find out what the real accuracy is and how the different errors have entered into the measurements. A study of errors is important in finding ways to reduce them and also to estimate the reliability of the final result.

Errors can be classified as:

1) Gross Errors
2) Systematic Errors
3) Random Errors

1) Gross Errors: are largely due to human errors - mistakes in reading and recording of data, incorrect adjustment and improper application of instruments and mistakes in calculations.

These errors can only be avoided by taking care in reading and recording the measured data correctly.

CULTIVATE THE GOOD HABITS OF:

(a) SELECTING THE PROPER SCALE AND RECORDING THE SCALE USED IN MEASUREMENTS.

(b) ADJUSTING THE POINTER OF INSTRUMENT TO ZERO DEFLECTION BEFORE THE START OF THE EXPERIMENT.

(c) DOUBLE CHECKING THE MEASUREMENTS BY YOU AND/OR YOUR PARTNER.

2) Systematic Errors:

(a) Instrumental Errors: are errors inherent in the instruments because of the tolerances and limits in the electrical and mechanical designs. Instrument errors can also be introduced due to misuse and loading effects of instruments. For example: a well-calibrated voltmeter may read erroneously when connected across two points in a high resistance circuit.

The above errors may be avoided by selecting a suitable instrument for the particular application, applying the proper correction factors once the amount of error is known.

REMEMBER: CARELESS OR UNINFORMED USE OF AN INSTRUMENT MAY DO PERMANENT DAMAGE AS A RESULT OF OVER-LOADING AND OVERHEATING OF THE INSTRUMENT.

(b) Environmental Errors: are errors due to external conditions to the measuring instrument due to changes in temperature, humidity, pressure or of magnetic or electrostatic fields. Corrective measures include the air-conditioning, use of magnetic shields, hermetically sealing certain components in the instrument etc.

(c) Observational Errors: are errors introduced by the observer. An observer may tend to read higher (or lower) than the correct value, possibly because of his reading angle and failure to avoid parallax.

3) Random Errors: are errors due to unknown causes and occur even when all systematic errors have been taken into account. They become important in high-accuracy work. These unknown errors are probably caused by a large number of small, variable effects so that their cumulative effects may be negligible in some cases or have a large net positive or negative error introduced. The only way to offset these errors is by increasing the number of readings and using statistical analysis to obtain the best approximation of the true value.

NOTE: Most of the errors discussed can be avoided by the experimenter by proper usage of the instruments and in reading the instruments as accurately as possible.

It is important to discuss and comment on the accuracy of the reading obtained during the experimentation.

© 1987 by Prentice-Hall, Inc., A Division of Simon & Schuster, Englewood Cliffs, N.J. 07632. All rights reserved. Printed in the United States of America.

Experiment 1

SIMPLE DC CIRCUIT

Required Reading: Text, section 2.2 to 2.5

1.1 Objective:

- To become familiar with the use of meters.
- To plot and understand the I-V characteristics and power hyperbola contours of linear and non-linear resistors.
- To verify Ohm's law.

1.2 Prelab Assignment:

Consider the circuits shown in Fig. 1.1.

The I-V relationships for the load are as follows:

Load # 1 .... $I_1 = 2 \times 10^{-4} V_1$

Load # 2 .... $I_2 = 10^{-4} V_2$

Load # 3 .... $I_3 = 10^{-14} e^{+40V_3}$

© 1987 by Prentice-Hall, Inc., A Division of Simon & Schuster, Englewood Cliffs, N.J. 07632. All rights reserved. Printed in the United States of America.

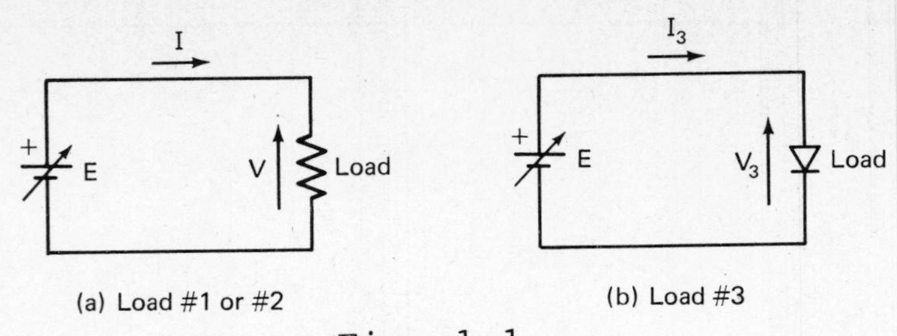

Fig. 1.1

(1) Determine the currents through loads 1 and 2 for the following values of voltages:

0, 5, 10, 15, 20, 25 volts

and plot the I-V characteristics of the loads on Graph 1.1.

(2) Determine the locus of points on the I-V characteristics of Graph 1.1 that represent a power dissipation of 50 mW. [Hint: calculate $I = \frac{(50 \text{ mW})}{V}$ mA for the values of the voltages suggested in (1)].

(3) Plot the I-V characteristics of load #3 on Graph 1.2 for a voltage range of 0 to 0.75 volts. Calculate, using the given I-V relationship, the resistance values at the following voltages and currents:

$V = 0.5$ volts, $V = 0.7$ volts,

$I = 1$ mA, $I = 10$ mA

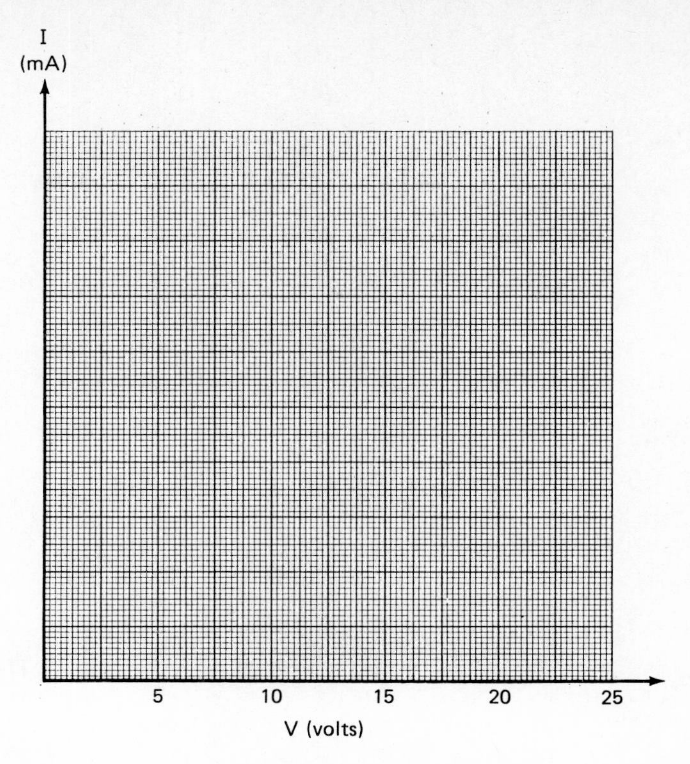

Graph 1.1

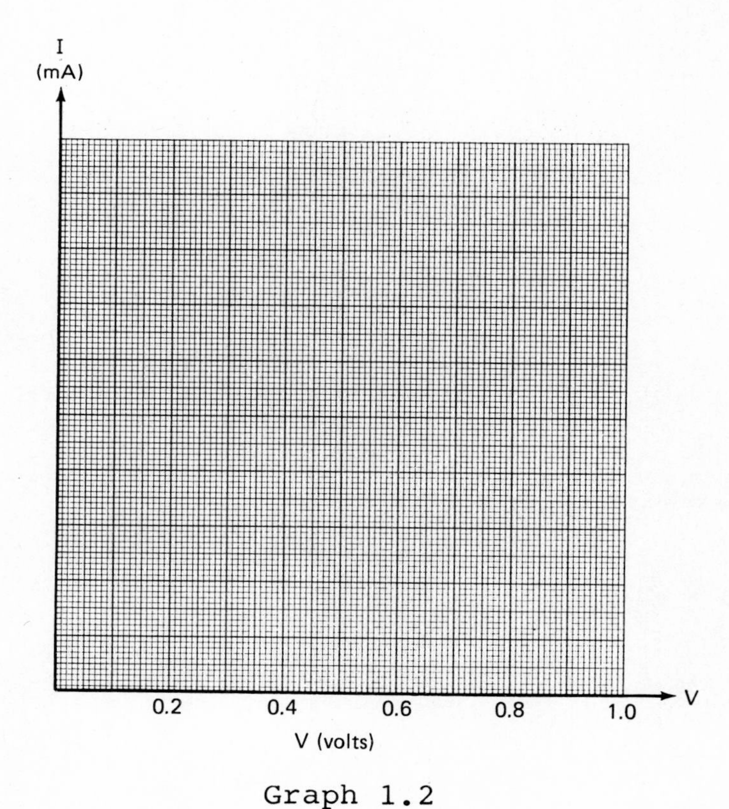

Graph 1.2

© 1987 by Prentice-Hall, Inc., A Division of Simon & Schuster, Englewood Cliffs, N.J. 07632. All rights reserved. Printed in the United States of America.

Prelab Work Space:

1.3 Equipment:

ITEM	MANUFACTURER AND MODEL NO.	LAB. SERIAL NO
DC Power Supply		
VOM		
DMM or milliammeter		

Resistors: One 4.7 kΩ and 10 kΩ .

1.4 Procedure:

A. I-V Characteristics:

(1) Connect the circuit shown in Fig 1.2 with R = 4.7 kΩ .

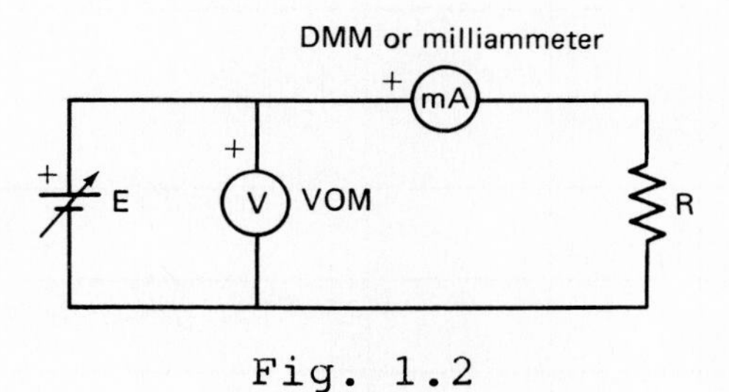

Fig. 1.2

(2) Vary the applied voltage (E) from 0 to 25 volts. Measure the voltage (V) and Current (I) values and record the results in Table 1.1.

© 1987 by Prentice-Hall, Inc., A Division of Simon & Schuster, Englewood Cliffs, N.J. 07632. All rights reserved. Printed in the United States of America.

(3) Plot the I-V characteristic of the 4.7 kΩ resistor on Graph 1.1 using the experimental results.

(4) Repeat steps 1 and 2 with R = 10 kΩ instead of 4.7 kΩ . Record your results in Table 1.1 and plot the I-V characteristic on Graph 1.1.

Table 1.1

R (kΩ) -nominal	VOLTMETER (V)		MILLIAMMETER (I)		OHM'S LAW $R=\frac{V}{I}$ kΩ	OHMMETER MEASUREMENT (kΩ)
	Range	Reading (Volts)	Range	Reading (mA)		
4.7		5				
		10				
		15				
		20				
		25				
10		5				
		10				
		15				
		20				
		25				

B. Resistance Measurements:

(1) Set the VOM as an ohmmeter and select the range-selector switch to X1 .

(2) Short-circuit the ohmmeter terminals and the meter should now show zero ohms or full-scale deflection on the current scale. If the pointer is not at zero ohms, adjust the potentiometer "ZERO" till the pointer is at zero. The meter is now ready to be used as an ohmmeter. (If the Ohm-range selector is switched to another range, the zeroing procedure should be followed for that range before using the meter as an ohmmeter.)

(3) Remove the short-circuit and connect the resistor with the unknown value across the ohmmeter terminals; the meter now indicates the ohmic value of the resistor. Using the above procedure, measure the ohmic values of the 4.7 kΩ and the 10 kΩ resistors noting that the R-scale is counterclockwise and non-linear. Record these values in Table 1.1.

(4) If a DMM is used to measure the unknown resistance value, set the meter on the

© 1987 by Prentice-Hall, Inc., A Division of Simon & Schuster, Englewood Cliffs, N.J. 07632. All rights reserved. Printed in the United States of America.

"resistance" mode of operation with the proper range and measure the resistance. Record the results in Table 1.1.

1.5 Comments and Conclusions:

1. Determine the average value of the resistance from the experimental I-V plot on Graph 1.1.

   R (from I-V graph) = ______ (nominal 4.7 kΩ resistor)

   = ______ (nominal 10 kΩ resistor)

2. How do the above experimental values compare with the nominal ratings and the measured values using the ohmmeter? Calculate the percent errors using the relationship:

$$\% \text{ Error} = \frac{[R(\text{measured}) - R(\text{nominal})]}{R(\text{nominal})} \times 100$$

3. How is Ohm's law confirmed in this experiment? Explain.

4. What is the relationship between the magnitude of the resistance and the slope of the I-V characteristic?

5. Draw the I-V characteristics on Graph 1.1 of a resistor with:

   a) R = 0 ohms

   b) R = Infinite Value

6. What are the possible reasons for any errors or deviations in the results from the theoretical or nominal values? Explain.

© 1987 by Prentice-Hall, Inc., A Division of Simon & Schuster, Englewood Cliffs, N.J. 07632. All rights reserved. Printed in the United States of America.

Experiment 2

SERIES DC CIRCUIT

Required Reading: Text, section 3.4

2.1 Objective:

- To verify Kirchhoff's voltage law and other properties of a simple series DC circuit.

2.2 Prelab Assignment:

(1) Consider the circuit shown in Fig. 2.1. The terminal voltage of the power supply is maintained at 25V.

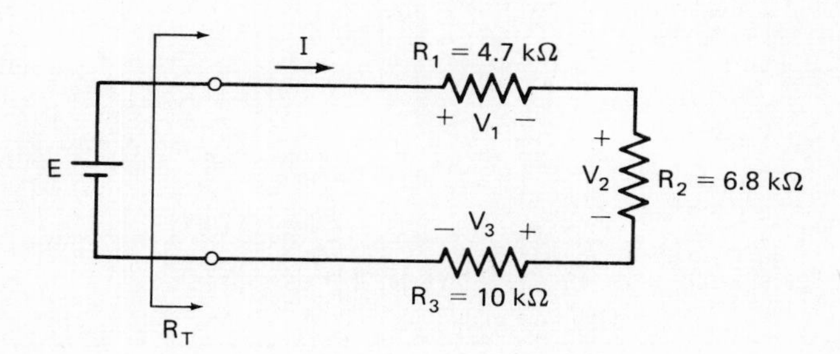

Fig. 2.1

© 1987 by Prentice-Hall, Inc., A Division of Simon & Schuster, Englewood Cliffs, N.J. 07632. All rights reserved. Printed in the United States of America.

Calculate:

(a) R_{Total}, I, $\frac{V_1}{E}$, $\frac{V_2}{V_3}$, $\frac{P_1}{P_T}$, $\frac{P_2}{P_3}$

(b) the maximum value of voltage that can be applied to the circuit without exceeding any of the component's power rating assuming that each resistor has a power rating of 1/2 watt.

Which resistor in this circuit is considered as the weakest link? Explain.

(2) Plot the I-V characteristics of R_1, R_2, R_3 and R_T on Graph 2.1.

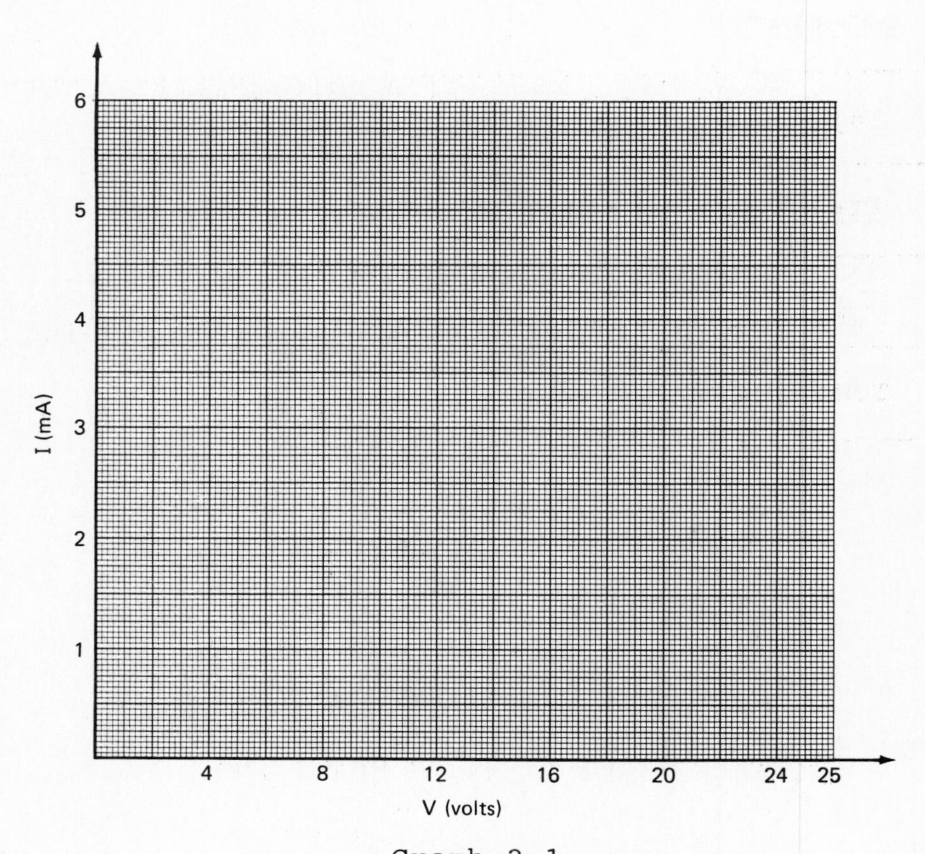

Graph 2.1

(3) What are the effects with respect to the voltage across, current through and power dissipation of each resistor, if the resistance of R_3 is increased? Record your answers in Table 2.1 using the following simple conventions:

'increase' ↑

'decrease' ↓

'no change' =

Table 2.1

COMP	R	V	I	P
R_1	=			
R_2	=			
R_3	↑			
TOTAL		=		

© 1987 by Prentice-Hall, Inc., A Division of Simon & Schuster, Englewood Cliffs, N.J. 07632. All rights reserved. Printed in the United States of America.

Prelab Work Space:

Prelab Work Space:

© 1987 by Prentice-Hall, Inc., A Division of Simon & Schuster, Englewood Cliffs, N.J. 07632. All rights reserved. Printed in the United States of America.

2.3 **Equipment:**

ITEM	MANUFACTURER AND MODEL NO.	LAB. SERIAL NO
DC Power Supply		
VOM		
DMM or DC milliammeter		

Resistors: One 4.7 kΩ , 6.8 kΩ , 10 kΩ , 15kΩ

2.4 **Procedure:**

(1) Connect the circuit shown in Fig. 2.2. The terminal voltage of the power supply is set at 25V.

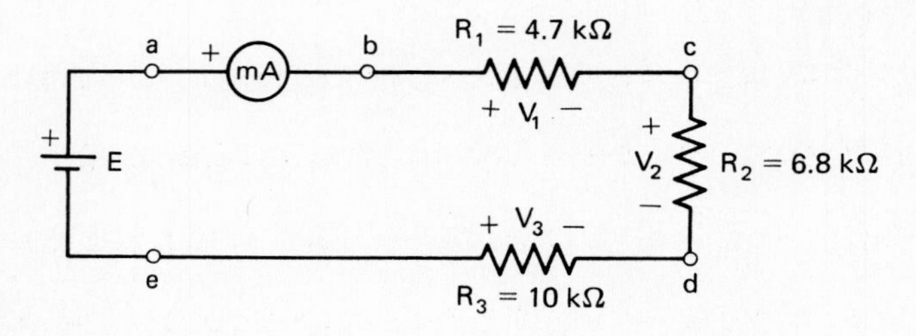

Fig. 2.2

(2) Measure the applied voltage (E), V_1, V_2 and V_3 and the current supplied from source (I_T). Record the results in Table 2.2.

(3) Disconnect the milliammeter from its present location and connect the meter at the proper locations in the above circuit to measure the

currents through R_1, R_2, and R_3 respectively. Record your results in Table 2.2.

Table 2.2

COMPONENT	VOLTMETER		MILLIAMMETER		R (calculated) $=(\frac{volts}{mA})$ kΩ
	Range	Reading (Volts)	Range	Reading (mA)	
DC POWER SUPPLY		25			R_T=
R_1					R_1=
R_2					R_2=
R_3					R_3=

(4) Calculate the values of R_T, R_1, R_2 and R_3 from the voltage and current measurements. Record these values in Table 2.2. Plot the I-V characteristics, using the above readings, of all the resistors on Graph 2.1.

(5) Compare the following quantities in Table 2.3 using the experimental values in Table 2.2:

© 1987 by Prentice-Hall, Inc., A Division of Simon & Schuster, Englewood Cliffs, N.J. 07632. All rights reserved. Printed in the United States of America.

Table 2.3

$V_1 + V_2 + V_3 =$ volts	$E =$ volts
$R_1 + R_2 + R_3 =$ kΩ	$R_T =$ kΩ
$\frac{V_1}{V_2} =$	$\frac{R_1}{R_2} =$
$\frac{V_3}{E} =$	$\frac{R_3}{R_T} =$

(6) Replace R_3 in Fig 2.2 by a 15 kΩ resistor. Measure I_T, V_1, V_2, V_3 and E and record these values in Table 2.4. Calculate the power dissipated in each resistor and total power.

Table 2.4

E (V)	V_1 (V)	V_2 (V)	V_3 (V)	I_T (mA)	POWER = (VI_T) mW			
					P_1	P_2	P_3	P_T

How does the increase in the magnitude of R_3 affect the voltage across, current through and power dissipation of each resistor in the circuit? (Compare the results of Table 2.4 with the results in Table 2.2). Record these results in Table 2.1, using the ↑ , ↓ , = convention. [For clarity, use a different color pencil or pen to emphasize the experimental results.]

2.5 Comments and Conclusions:

1. Summarize the properties of a DC series circuit. Are these properties supported by your experimental results? Use the various data obtained in the experiments to justify the statements.

© 1987 by Prentice-Hall, Inc., A Division of Simon & Schuster, Englewood Cliffs, N.J. 07632. All rights reserved. Printed in the United States of America.

2. What are the possible causes for any deviations in the experimental results from the theoretical or nominal values? Explain.

3. Is the Loading effect of VOM as the voltmeter significant in measuring the voltages in the circuit of Fig. 2.2? Explain.

4. What are the effects on I_T, V_1, V_2 and V_3 if R_1, for some reason, is:

   (a) open-circuited;

   (b) short-circuited?

Experiment 3

PARALLEL DC CIRCUIT

Required Reading: Text, section 3.5

3.1 **Objective:**

To verify Kirchhoff's current law and other properties of a parallel circuit.

3.2 **Prelab Assignment:**

(1) Consider the circuit shown in Fig. 3.1. The terminal voltage of the power supply is maintained at 25V.

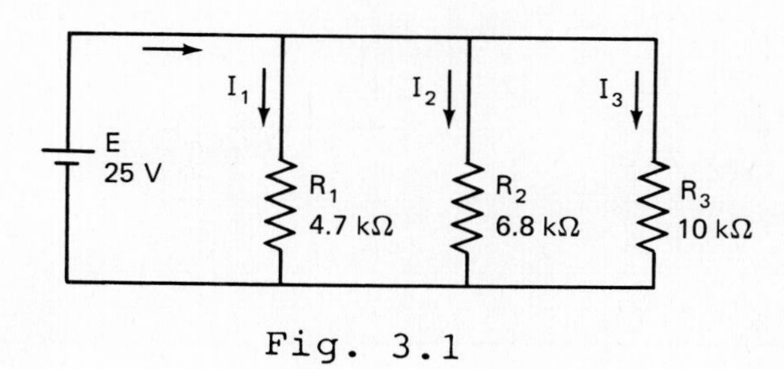

Fig. 3.1

© 1987 by Prentice-Hall, Inc., A Division of Simon & Schuster, Englewood Cliffs, N.J. 07632. All rights reserved. Printed in the United States of America.

Calculate:

(a) $R_T, G_T, I_T, \frac{I_1}{I_T}, \frac{I_2}{I_3}, \frac{G_1}{G_T}, \frac{G_2}{G_3}, \frac{P_1}{P_T}, \frac{P_2}{P_3}$

(b) the maximum value of I_T that can be supplied to the circuit without exceeding any of the component's power rating, assuming that each resistor has a power rating of 1/2 watt.
Which resistor in this circuit is considered as the weakest link? Explain.

(2) Plot the I-V characteristics of R_1, R_2, R_3 and R_T on Graph 3.1.

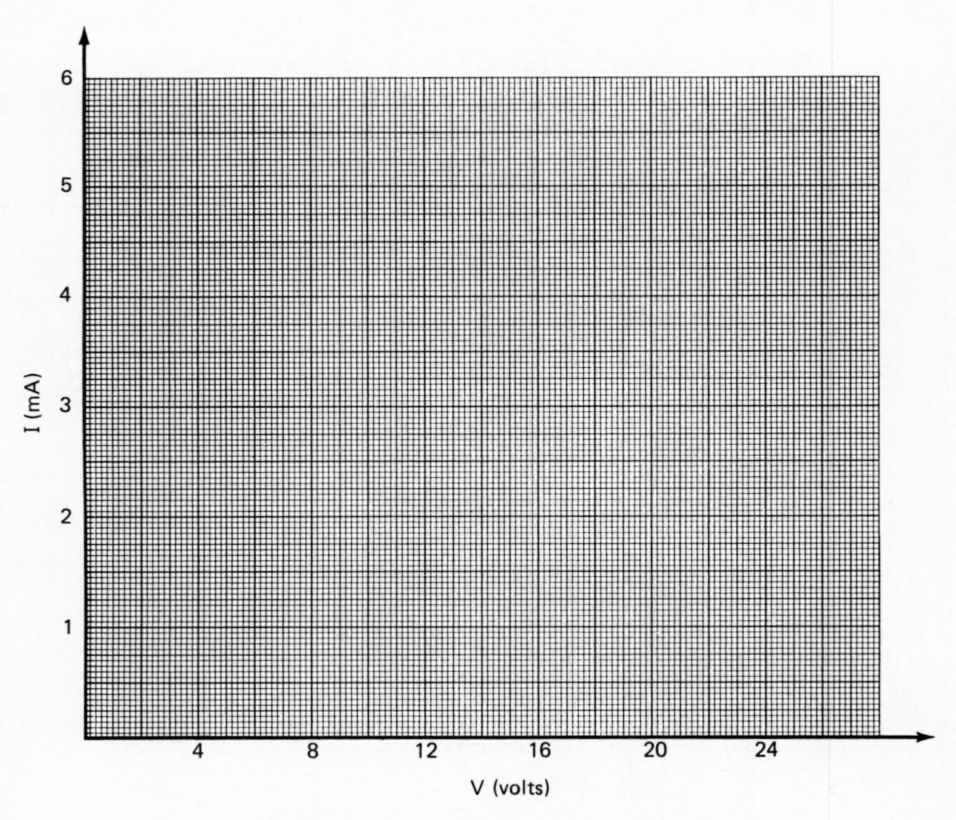

Graph 3.1

(3) What are the effects with respect to the voltage across, current through and power dissipation of each resistor, if the resistance of R_3 is increased? Record your answers in Table 3.1 using the following simple conventions:

'increase' ↑

'decrease' ↓

'no change' =

Table 3.1

COMPONENT	R	V	I	P
R_1	=			
R_2	=			
R_3	↑			
TOTAL		=		

© 1987 by Prentice-Hall, Inc., A Division of Simon & Schuster, Englewood Cliffs, N.J. 07632. All rights reserved. Printed in the United States of America.

Prelab Work Space:

Prelab Work Space:

© 1987 by Prentice-Hall, Inc., A Division of Simon & Schuster, Englewood Cliffs, N.J. 07632. All rights reserved. Printed in the United States of America.

3.3 **Equipment:**

ITEM	MANUFACTURER AND MODEL NO.	LAB. SERIAL NO
DC Power Supply		
VOM		
DMM or DC milliammeter		

Resistors: One 4.7 kΩ , 6.8 kΩ , 10 kΩ , 15kΩ

3.4 **Procedure:**

(1) Connect the circuit shown in Fig 3.2.

The terminal voltage of the power supply is set at 25V.

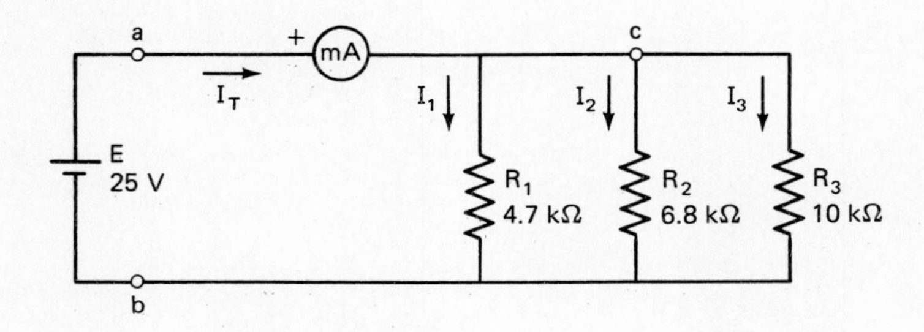

Fig. 3.2

(2) Measure the total current supplied from source (I_T). Using the VOM or DMM, check and verify that the voltage across each resistor is 25V. Record your results in Table 3.2.

Table 4.1

(a) R_3: decreased

COMPONENT	R	V	I	P
R_1	=			
R_2	=			
R_3	↓			
R_4	=			
R_5	=			
TOTAL		=		

Table 4.1

(b) R_1: increased

COMPONENT	R	V	I	P
R_1	↑			
R_2	=			
R_3	=			
R_4	=			
R_5	=			
TOTAL		=		

© 1987 by Prentice-Hall, Inc., A Division of Simon & Schuster, Englewood Cliffs, N.J. 07632. All rights reserved. Printed in the United States of America.

Prelab Work Space:

4.3 Equipment:

ITEM	MANUFACTURER AND MODEL NO.	LAB. SERIAL NO
DC Power Supply		
VOM		
DMM or DC milliammeter		
Decade Resistance Box		

Resistors: One 4.7 kΩ , 6.8 kΩ , 10 kΩ , 15kΩ and 22 kΩ .

© 1987 by Prentice-Hall, Inc., A Division of Simon & Schuster, Englewood Cliffs, N.J. 07632. All rights reserved. Printed in the United States of America.

4.4 **Procedure:**

A. Verification of KVL and KCL:

(1) Connect the circuit as shown in Fig. 4.2. Set the output voltage of the power supply at 25V.

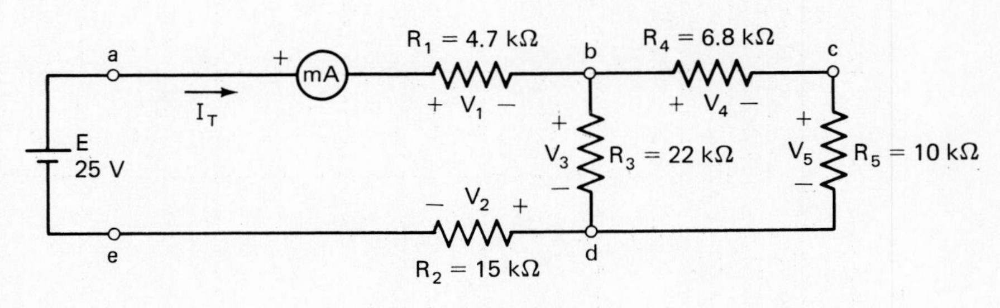

Fig. 4.2

(2) Measure the total current (I_T) supplied from the source; the voltage drops V_1, V_2, V_3, V_4, V_5 and the applied voltage E. Record your results in Table 4.2.

(3) Disconnect the ammeter from its present location and reconnect it back in the circuit at various locations in order to measure I_2, I_3, I_4 and I_5 successively. Record these results in Table 4.2.

Table 4.2

		E (V)	V_1 (V)	V_2 (V)	V_3 (V)	V_4 (V)	V_5 (V)
VOLTS	CALC (V) -Prelab						
	EXPT. (V)						
		I_T	I_1	I_2	I_3	I_4	I_5
CURRENTS	CALC. (mA) -Prelab						
	EXPT. (mA)						
		R_T	R_1	R_2	R_3	R_4	R_5
RESISTANCE	-Prelab (kΩ)		4.7	15	22	6.8	10
	EXPT. (kΩ)						

(4) Replace R_3 in the circuit of Fig. 4.2 by a decade resistance box. Set the resistance of the box at 5 kΩ. Measure E, I_T, V_1, V_2, V_3, V_4, V_5, I_1, I_2, I_3, I_4 and I_5. Record these results in Table 4.3.

© 1987 by Prentice-Hall, Inc., A Division of Simon & Schuster, Englewood Cliffs, N.J. 07632. All rights reserved. Printed in the United States of America.

Table 4.3

R_3 = 5 kΩ

E (V)	V_1 (V)	V_2 (V)	V_3 (V)	V_4 (V)	V_5 (V)	I_T (mA)	I_1 (mA)	I_2 (mA)	I_3 (mA)	I_4 (mA)	I_5 (mA)
25											

(5) Return R_3 back to the circuit (R_3 = 22 kΩ) and now replace R_1 by a decade resistance box set to 25 kΩ . Measure all the voltages and currents in the circuit. Record the results in Table 4.4.

Table 4.4

R_1 = 25 kΩ

E (V)	V_1 (V)	V_2 (V)	V_3 (V)	V_4 (V)	V_5 (V)	I_T (mA)	I_1 (mA)	I_2 (mA)	I_3 (mA)	I_4 (mA)	I_5 (mA)
25											

(6) Compare the results in Tables 4.3 and 4.4 with the corresponding quantities in Table 4.2 and superimpose the observations in Table 4.1, using ' ↑ , ↓ , = ' convention.

B. <u>Concept of circuit 'ground':</u>

(1) With node 'b' as circuit 'ground', measure V_a, V_c, V_d, V_e. [Hint: one terminal of the voltmeter is connected to node 'b'.] Record your results in Table 4.5.

(2) With node 'd' as circuit 'ground' (instead of node 'b'), measure V_a, V_b, V_c, V_e. Record your results in Table 4.5.

Table 4.5

	'b' circuit ground		'd' circuit ground	
Potential	CALC Prelab (V)	EXPT. (V)	CALC Prelab (V)	EXPT. (V)
V_a				
V_b				
V_c				
V_d				
V_e				

© 1987 by Prentice-Hall, Inc., A Division of Simon & Schuster, Englewood Cliffs, N.J. 07632. All rights reserved. Printed in the United States of America.

4.5 Comments and Conclusions:

1. Apply Kirchhoff's voltage and current laws to the series-parallel circuit in Fig. 4.2. Were these laws supported by your experimental results? Explain.

2. What are the possible causes for any deviations in the experimental results from the theoretical values?

3. What is a circuit 'ground' and why is there a necessity of a circuit 'ground' in some circuits?

4. Explain the effects of moving the circuit ground from 'b' to 'd' on:

   (a) currents in the circuit,

© 1987 by Prentice-Hall, Inc., A Division of Simon & Schuster, Englewood Cliffs, N.J. 07632. All rights reserved. Printed in the United States of America.

(b) voltage drops across the resistors,

(c) potentials of various nodes.

5. During experimentation with the node 'b' as circuit ground, node 'd' was accidentally grounded. What are the effects on currents and voltages in the circuit due to such an incident? Draw the equivalent circuit of Fig. 4.2 for this condition.

Experiment 5

WHEATSTONE BRIDGE

Required Reading: Text, section 3.7

5.1 Objective:

To understand the principle and performance of the Wheatstone bridge.

5.2 Prelab Assignment:

A. Simple Methods of Measuring Unknown Resistance:

Explain the procedure with the necessary circuit diagrams that can be used in the laboratory to measure an unknown resistance given the following sets of equipment only.

(1) Set # 1:

one DC power supply

one voltmeter

one ammeter

© 1987 by Prentice-Hall, Inc., A Division of Simon & Schuster, Englewood Cliffs, N.J. 07632. All rights reserved. Printed in the United States of America.

(2) Set # 2:

one DC power supply

one voltmeter

one decade resistance box

Explain the limitations of each of the above methods.

B. Wheatstone bridge:

The circuit of Fig. 5.1 shows a simple schematic of the Wheatstone bridge. R_1 and R_2 can each have three possible settings: $10\ \Omega$, $100\ \Omega$ and $1000\ \Omega$. R_3 is a decade resistance box with a maximum possible value of $9999\ \Omega$. R_x is the unknown resistor and may assume any value from $1\ m\Omega$ to a high value of $1\ M\Omega$.

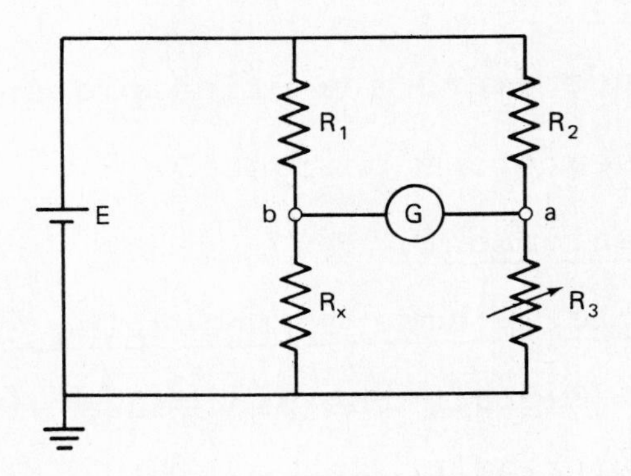

Fig. 5.1

1(a) Determine the values of R_1, R_2, R_3 and R_x which result in the potentials V_a to be as large as possible and V_b to be as small as possible. What is the direction of the current flow through the galvanometer (G): from 'a' towards 'b' or 'b' to 'a'?

(b) If the unknown resistor R_x is variable and is gradually increased, how would this change affect the magnitudes of V_a and V_b? Determine the value of R_x (max) at which V_a and V_b are equal with the settings of R_1, R_2 and R_3 as in (a).

(c) If R_x is larger than R_x (max), what would be the direction of current flow through the galvanometer? Could the bridge be balanced under this condition with the settings of R_1, R_2 and R_3 available on the bridge? Explain.

(2) Suppose that an unknown resistance (R_x) is in the range of 0 to 99 Ω and is to be measured using the above bridge. What are the best settings of R_1 and R_2 that can achieve the optimum accuracy of a measurement of R_x? [Note: R_3 has a maximum value consisting of four digits.]

© 1987 by Prentice-Hall, Inc., A Division of Simon & Schuster, Englewood Cliffs, N.J. 07632. All rights reserved. Printed in the United States of America.

Prelab Work Space:

5.3 Equipment:

ITEM	MANUFACTURER AND MODEL NO.	LAB. SERIAL NO
DC Power Supply		
VOM		
DMM or DC milliammeter		
Commercial Wheatstone Bridge		
Decade Resistance Boxes: Three Nos.		
Galvanometer or Null Detector		

Resistors: One 47 Ω , 470 Ω , 6.8 kΩ

5.4 Procedure:

A. Simple Methods:

1(a) Assume that 470 Ω resistor is the unknown resistance (R_X). Using the equipment of Set #1 (Prelab), measure the unknown resistance. Set the output voltage of DC power supply at 10V. Read the milliammeter and voltmeter readings as accurately as possible. Record the results in Table 5.1.

1(b) Repeat step (a) with R_X = 6.8 k Ω . Record the results in Table 5.1.

© 1987 by Prentice-Hall, Inc., A Division of Simon & Schuster, Englewood Cliffs, N.J. 07632. All rights reserved. Printed in the United States of America.

Table 5.1

R (nominal)	Voltmeter reading (V)	Ammeter reading (mA)	R_x (calculated)
470 Ω			
6.8 kΩ			

(2) Use the equipment of Set # 2 (Prelab). With the DC power supply set at 10V, measure the unknown resistance, assuming 470 Ω and 6.8 kΩ resistors as unknown values. Record the values in Table 5.2.

Table 5.2

R (nominal)	R (Decade box)	R_x
470 Ω		
6.8 kΩ		

B. Wheatstone bridge:

(1) Connect the circuit of Fig. 5.2.

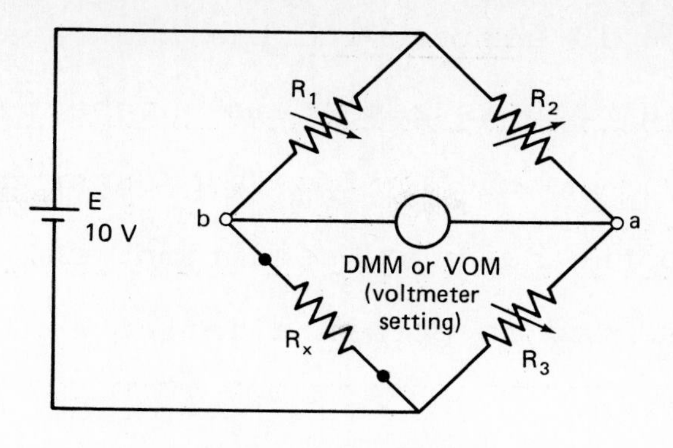

Fig. 5.2

Set E = 10V, R_3 at its <u>maximum</u> value and R_x = 470 Ω. Set the VOM at a higher voltage scale. Select the values of R_1 = 100 Ω and R_2 = 1000 Ω .

Balance the bridge by reducing R_3 till the voltage V_{ab} is zero or as close to zero as possible. Accurate balance of the bridge can be achieved by reducing the voltage scales of VOM and adjusting R_3 further till the voltmeter reads zero. Record the value of R_3 and calculate R_x.

(2) Repeat the measurements for the various values of R_1, R_2 and R_x shown in Table 5.3. <u>Before each new setting of R_1 and R_2 are selected, R_3 should be adjusted back to its maximum value and the VOM on a higher scale.</u>

Caution: If the VOM is used in the current mode (a current detector) or a galvanometer is used (instead of VOM) between the points a-b, it is advisable to use a potentiometer or a variable

© 1987 by Prentice-Hall, Inc., A Division of Simon & Schuster, Englewood Cliffs, N.J. 07632. All rights reserved. Printed in the United States of America.

resistance in series with the instrument to protect it. Resistance is set at its highest value initially. As the bridge is brought closer and closer to the balanced condition, this resistance can be reduced to obtain a better sensitivity of the detector.

Table 5.3

R_x (nominal)	R_1 (Ω)	R_2 (Ω)	R_3 (Ω)	$R_x = (\frac{R_1}{R_2}) R_3$
	100	1000		
(a) 470 Ω	100	100		
	1000	100		
	100	1000		
(b) 6800 Ω	100	100		
	1000	100		
	100	1000		
(c) 47 Ω	100	100		
	1000	100		

C. <u>Commercial Wheatstone bridge:</u>

In commercial bridges, R_1 and R_2 are variable in decades so that the ratio of R_1/R_2 is an integral multiplier or decimal and is normally called 'multiplier ratio' or 'ratio arm'. R_3 is a decade resistor (or a slide wire rheostat). Null detector is usually a sensitive galvanometer with a push-button switch and a potentiometer in series with the detector.

(1) Record all the ratios of (R_1/R_2) available on the bridge and also the maximum-value of R_3 obtainable. This data would enable you to decide on which of the ratios of (R_1/R_2) are useful in measuring R_x.

(2) With R_x set at 470 Ω, measure the value of R_x for all the possible ratios of (R_1/R_2). Record the results in Table 5.4.

(3) Repeat (2) for R_x = 6800 Ω and R_x = 47 Ω. Record the results in Table 5.4.

© 1987 by Prentice-Hall, Inc., A Division of Simon & Schuster, Englewood Cliffs, N.J. 07632. All rights reserved. Printed in the United States of America.

Table 5.4

R_x (nominal)	(R_1/R_2)	R_3	$R_x = (\frac{R_1}{R_2}) R_3$	% Error
(a) 470 Ω				
(b) 6800 Ω				
(c) 47 Ω				

$$\% \text{ Error} = \frac{[R_x(\text{measured}) - R_x(\text{nominal})]}{R_x(\text{nominal})} \times 100$$

5.5 **Comments and Conclusions:**

1. What are the various methods used in the laboratory for measuring unknown resistance?

2. Which of the methods in (1) is the most accurate one? Explain.

3. What are the limitations in measuring R_X using methods in Part (A)?

© 1987 by Prentice-Hall, Inc., A Division of Simon & Schuster, Englewood Cliffs, N.J. 07632. All rights reserved. Printed in the United States of America.

4. What are the maximum and minimum values of R_x that can be measured in:

(a) circuit of Fig. 5.2,

(b) commercial Wheatstone bridge?

5. How can the Wheatstone bridge be used as a:

(a) Thermometer,

(b) Lie-detector?

Experiment 6

PRACTICAL DC SOURCES

Required Reading: Text, section 4.1 to 4.5

6.1 Objective:

- To find the parameters of a practical DC source and investigate the behaviour of a practical DC source under varying load conditions.
- To become familiar with the 'load line' concept.
- To understand the 'maximum power transfer' principle.

6.2 Prelab Assignment:

Consider a DC power supply with the following characteristics:

-terminal voltage is 35V on no-load condition and 30 volts when a load of 270 Ω is connected to the terminals:

© 1987 by Prentice-Hall, Inc., A Division of Simon & Schuster, Englewood Cliffs, N.J. 07632. All rights reserved. Printed in the United States of America.

-internal resistance is <u>unknown</u>.

(1) Calculate the internal resistance of the source.

(2) Plot the load line (regulation curve) of the power supply on Graph 6.1. Draw the V-I characteristic of the 270 Ω load on the same graph. Determine graphically the load current (I_L) and voltage (V_L) across the load.

(3) A variable resistance load (R_L) is connected across the power supply and varied over a range of 5 to 500 Ω in the steps shown in Table 6.1. Calculate the values of I_L, V_L, P_L, the total power supplied from the source and the efficiency of the power transfer for each setting of R_L. Record these results in Table 6.1.

Table 6.1

R_L (Ω)	5	20	40	45	50	60	100	200	500
I_L (mA)									
V_L (V)									
P_L (W)									
P_T (W)									
% Efficiency $=(P_L/P_T) \times 100$									

(4) Plot the graphs of the load current (I_L) versus R_L and the voltage across the load (V_L) versus R_L on Graph 6.2.

(5) Plot the graphs of the power dissipated in the load (P_L) and the power-transfer efficiency versus the load resistance (R_L) on Graph 6.3.

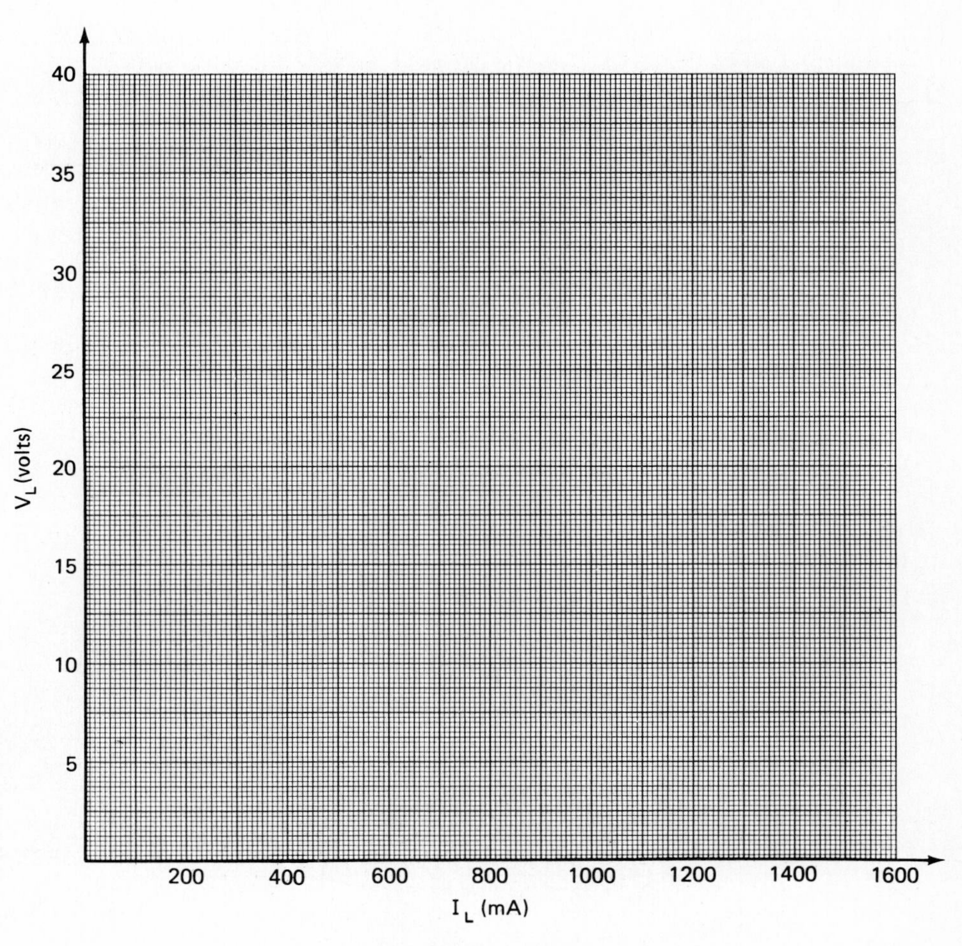

Graph 6.1

© 1987 by Prentice-Hall, Inc., A Division of Simon & Schuster, Englewood Cliffs, N.J. 07632. All rights reserved. Printed in the United States of America.

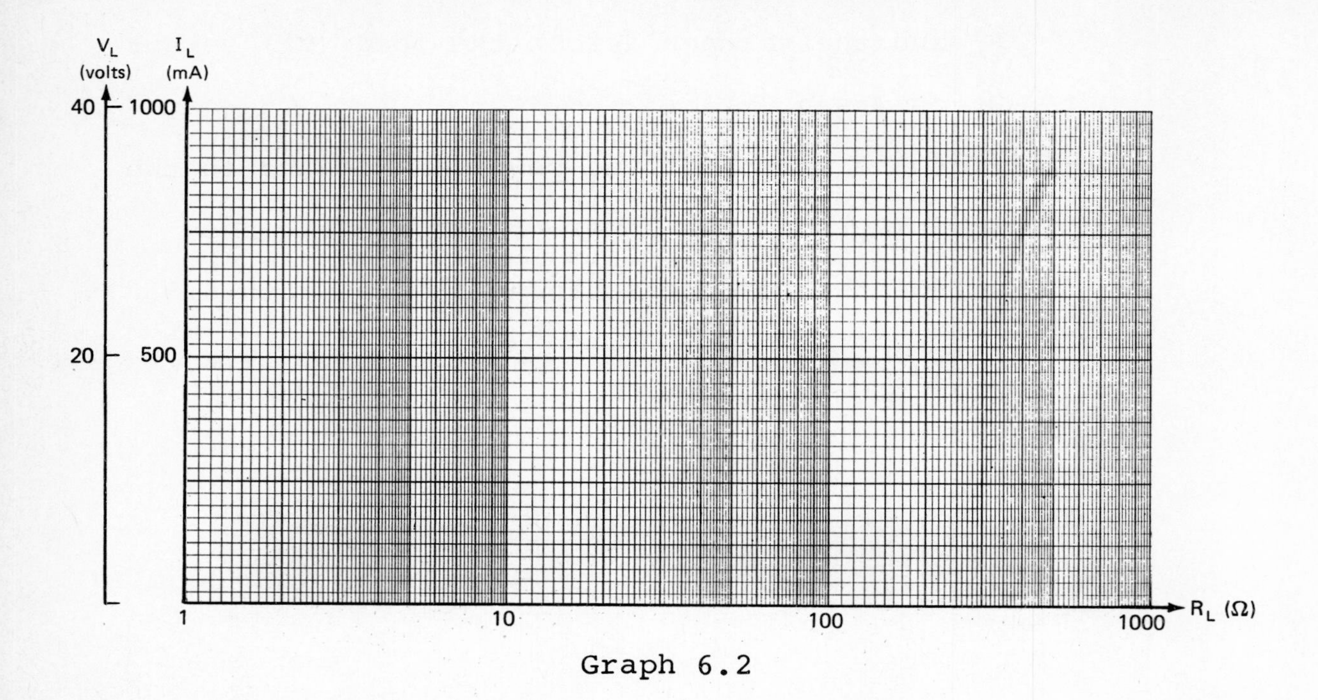

Graph 6.2

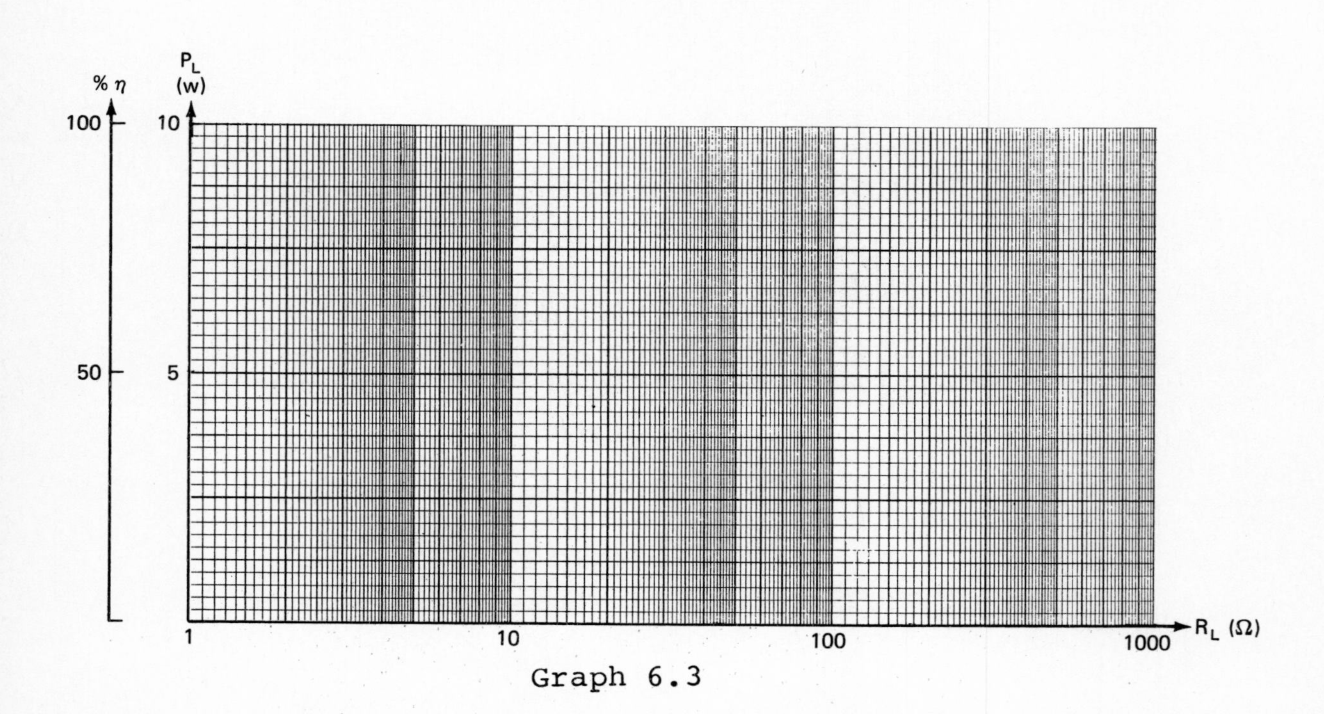

Graph 6.3

(2) Vary the load resistor by the values shown in Table 6.2 and for each setting, measure the current through the load (I_L) and the voltage across the load (V_L). Record the results in Table 6.2. Calculate P_L, P_T and efficiency of the power transfer for each value of R_L and record these values in Table 6.2.

Table 6.2

R_L (Ω)	5	20	40	45	50	60	100	200	500
I_L (mA)									
V_L (V)									
P_L (W)									
P_T (W)									
% Efficiency $=(P_L/P_T) \times 100$									

(3) Superimpose the V_L versus I_L graph on Graph 6.1. From the measure data, determine the actual internal resistance (R_{int}) of the power supply.

(4) Estimate the short-circuit current (I_{sc}) from the above graph.

© 1987 by Prentice-Hall, Inc., A Division of Simon & Schuster, Englewood Cliffs, N.J. 07632. All rights reserved. Printed in the United States of America.

(5) Plot the V-I characteristic of $R_L = 50\,\Omega$ on Graph 6.2 and determine graphically the voltage across and current through R_L. Compare this value with the measured value in Table 6.2.

(6) Superimpose the graphs of I_L versus R_L and V_L versus R_L on Graph 6.2; the graphs of P_L and efficiency versus R_L on Graph 6.3.

6.5 Comments and Conclusions:

1. What should be the value of the internal resistance relative to the load resistance of a good voltage source? Relatively large or relatively small. Explain.

2. What should be the value of the internal resistance relative to R_L of a good current source? Relatively large or relatively small. Explain.

3. For what value of R_L, is maximum power transferred to the load?

4. What is the value of the power transfer efficiency at the maximum power transfer condition?

5. Do you consider the 'DC power supply' used in this experiment a good voltage source? Explain.

© 1987 by Prentice-Hall, Inc., A Division of Simon & Schuster, Englewood Cliffs, N.J. 07632. All rights reserved. Printed in the United States of America.

6. If the load is a non-linear resistor whose V-I characteristic is given, explain how you would determine the current and voltage across the load when connected to the above DC supply.

7. Compare and comment on the experiment results with respect to the expected prelab results. Explain the reasons for deviations.

Experiment 7

VOLTAGE DIVIDERS

Required Reading: Text, section 4.6

7.1 Objective:

To become familiar with 'voltage divider' concepts.

7.2 Prelab Assignment:

A. Divider for a Variable Load:

Design a voltage divider to supply power to a varying load, using a DC power supply with an open-circuit voltage of 25V and an internal resistance of 45 Ω. The load voltage (V_L) must remain within the range of

$10V \leq V_L \leq 11$ volts, while the load (R_L) is allowed to vary over the range

$1\ k\Omega \leq R_L \leq 10\ k\Omega$.

© 1987 by Prentice-Hall, Inc., A Division of Simon & Schuster, Englewood Cliffs, N.J. 07632. All rights reserved. Printed in the United States of America.

B. Divider for Multiple Loads:

Design a voltage divider to supply the following loads:

3 mA at + 14 volts,

1 mA at - 10 volts.

The source has an open-circuit voltage of 30V and an internal resistance of 45 Ω. The supply current is limited to 10 mA.

Prelab Work Space:

Prelab Work Space:

© 1987 by Prentice-Hall, Inc., A Division of Simon & Schuster, Englewood Cliffs, N.J. 07632. All rights reserved. Printed in the United States of America.

Prelab Work Space:

7.3 Equipment:

ITEM	MANUFACTURER AND MODEL NO.	LAB. SERIAL NO
DC Power Supply		
VOM		
DMM or DC milliammeter		
Decade Resistance Box: 3 Nos.		

Resistors: One 4.7 kΩ and 10 kΩ .

7.4 Procedure:

A: Divider for a Variable Load:

(1) Connect the circuit as in Fig. 7.1 using your design of the voltage divider. Adjust the open-circuit voltage of a DC power supply to 25V. Use the three decade resistance boxes to represent the voltage divider and the variable load (R_L).

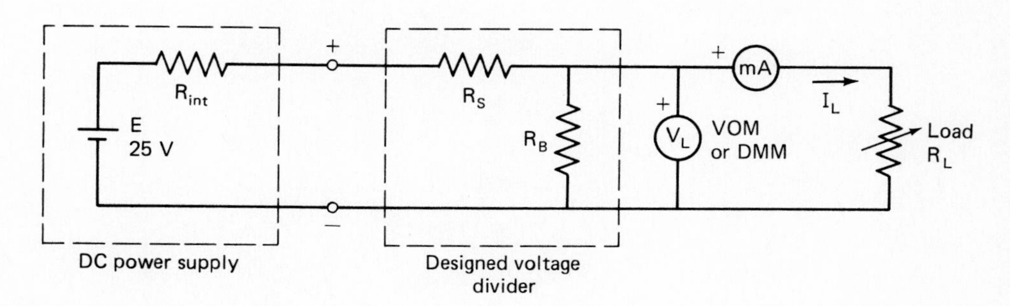

Fig. 7.1

© 1987 by Prentice-Hall, Inc., A Division of Simon & Schuster, Englewood Cliffs, N.J. 07632. All rights reserved. Printed in the United States of America.

The Voltage divider has a series-drop resistor (R_S) and a bleeder resistor (R_B). [If the internal resistance of the power supply is different from 45 Ω, adjust the series-drop resistor to account for the change in the internal resistance.]

(2) Set the values of R_L to the various values as suggested in Table 7.1 and measure I_L and V_L for each of these values. Record these results in the table.

Table 7.1

R_L (kΩ)	10	9	7	5	4	3	2	1
I_L (mA)								
V_L (V)								

(3) Plot the voltage regulation characteristics on Graph 7.1.

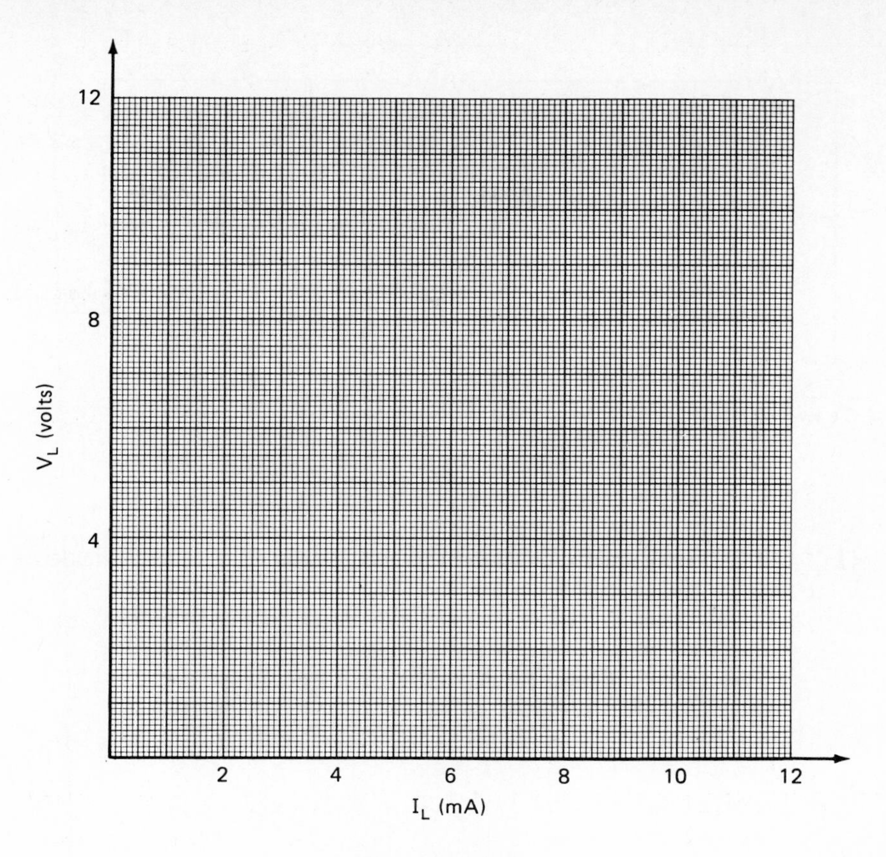

Graph 7.1

B. <u>Divider for Multiple Loads:</u>

(1) Connect the circuit as in Fig. 7.2 using your design of the voltage divider. Set the open-circuit voltage of a DC power supply to 30 volts. Use the three decade resistance boxes to represent the designed voltage divider. [If the internal resistance of the power supply is different from 45 Ω , the series-drop resistor must be adjusted correspondingly.]

© 1987 by Prentice-Hall, Inc., A Division of Simon & Schuster, Englewood Cliffs, N.J. 07632. All rights reserved. Printed in the United States of America.

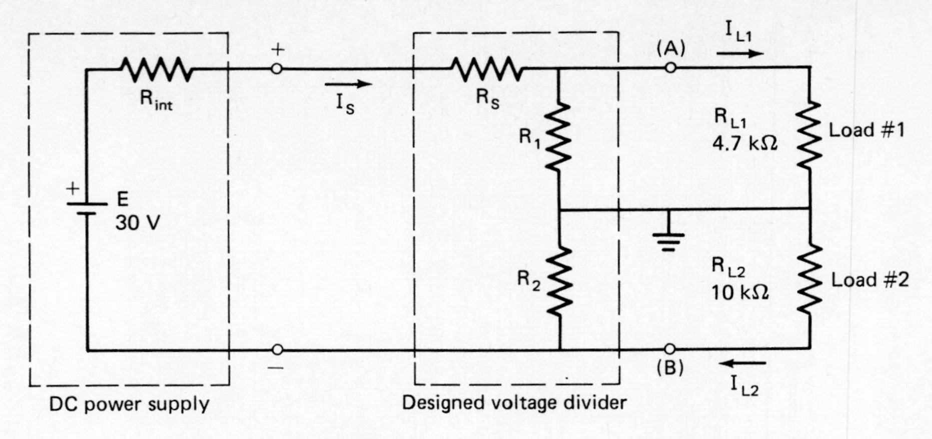

Fig. 7.2

(2) Use the milliammeter or DMM in the ammeter mode to measure the current supplied from the source (I_S) and the load currents I_{L1} and I_{L2}. Measure the potential V_A and V_B. Record your results in Table 7.2.

Table 7.2

V_A (V)	I_{L1} (mA)	P_{L1} (mW)	V_B (V)	I_{L2} (mA)	P_{L2} (mW)	I_S (mA)	P_T (mW)

(3) Calculate the efficiency of the power transfer from the source to the loads:

$$\% \text{ Efficiency} = \frac{(P_{L1} + P_{L2})}{P_T} \times 100$$

7.5 Comments and Conclusions:

1. What are the advantages and disadvantages of a simple voltage divider with the bleeder resistor compared with the divider without the bleeder?

2. Consider the voltage divider of Part A; Calculate:

   (a) Ratio of I_B/I_L at loads 1 kΩ and 10 kΩ ,

   (b) I_B as a percentage of the total current supplied from the source at the above loads.

   Comment on the advantage and disadvantage of the voltage divider, based on the data you calculated in (a) and (b).

© 1987 by Prentice-Hall, Inc., A Division of Simon & Schuster, Englewood Cliffs, N.J. 07632. All rights reserved. Printed in the United States of America.

3. How can the voltage regulation of the divider with the bleeder be improved? What compromise is necessary to improve the regulation as you suggested?

4. What is the effect on the current through and voltage drop across R_2, if R_1 in the voltage divider of Fig. 7.2 is shorted-out during the operation of the circuit? Assume that the power supply maintains a constant output voltage.

5. Compare and comment on the experimental results with respect to expected prelab calculations. Explain the possible reasons for any deviations.

© 1987 by Prentice-Hall, Inc., A Division of Simon & Schuster, Englewood Cliffs, N.J. 07632. All rights reserved. Printed in the United States of America.

Experiment 8

THEVENIN'S THEOREM

Required Reading: Text, sections 5.1, 5.2, 6.2 and 6.3.

8.1 Objective:

To verify Thevenin's theorem.

8.2 Prelab Assignment:

Consider the circuit shown in Fig. 8.1.

R_L is a variable load with the following resistance settings:

50 Ω , 100 Ω , 1.0 kΩ , 1.25 kΩ , 1.5 kΩ , 10 kΩ , 20 kΩ and 30 kΩ .

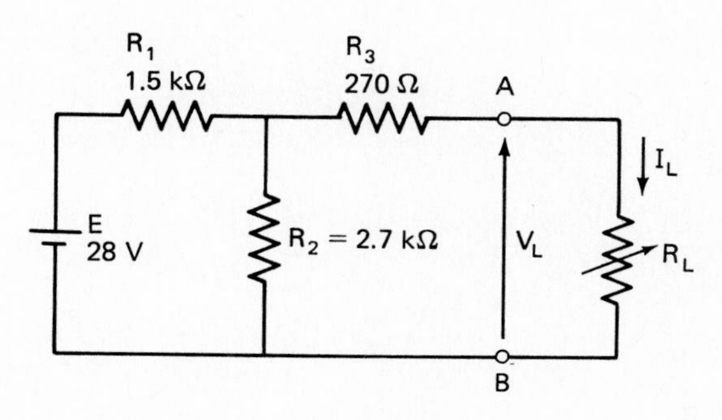

Fig. 8.1

© 1987 by Prentice-Hall, Inc., A Division of Simon & Schuster, Englewood Cliffs, N.J. 07632. All rights reserved. Printed in the United States of America.

A. Series-Parallel Circuit: Analysis by Loop or Mesh-Current Method

Using the "Loop current" analysis, determine the values of V_L and I_L for each setting of R_L. Record these values in Table 8.2.

B. Thevenin's Theorem:

(a) Find the Thevenin's equivalent circuit parameters as seen at the terminals A-B. Record your values in Table 8.1.

(b) Draw the Thevenin's equivalent circuit and using this circuit, calculate the voltage (V_L) across and current (I_L) through R_L for each setting of R_L. Record these values in Table 8.2.

Prelab Work Space:

© 1987 by Prentice-Hall, Inc., A Division of Simon & Schuster, Englewood Cliffs, N.J. 07632. All rights reserved. Printed in the United States of America.

Prelab Work Space:

8.3 Equipment:

ITEM	MANUFACTURER AND MODEL NO.	LAB. SERIAL NO
DC Power Supply		
VOM		
DMM or DC milliammeter		
Decade Resistance Box		

Resistors: One 270 Ω , 1.5 kΩ and 2.7 kΩ .

8.4 Procedure:

A: Series-Parallel Circuit:

(1) Connect the circuit as in Fig. 8.2.

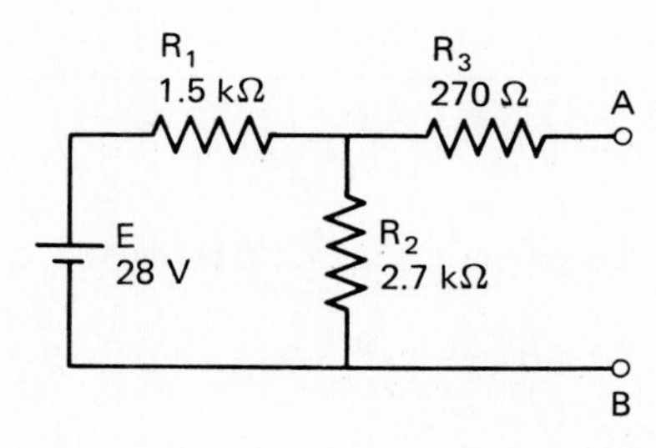

Fig. 8.2

(2) Connect a decade resistance box (R_L) across the A-B terminals with the necessary instruments to measure I_L and V_L. For each setting of R_L as shown in Table 8.2, measure I_L and V_L and record the results in Table 8.2.

© 1987 by Prentice-Hall, Inc., A Division of Simon & Schuster, Englewood Cliffs, N.J. 07632. All rights reserved. Printed in the United States of America.

B. Thevenin's Equivalent Circuit:

(1) Remove the load R_L. Measure the open-circuit voltage between the terminals A-B. Record this value of voltage as V_{Th} in Table 8.1.

(2) Connect a milliammeter between the terminals A-B and measure the short-circuit current (I_{AB}); calculate ($R_{Th} = V_{Th}/I_{AB}$). Record this result in Table 8.1.

Table 8.1

	V_{Th} (V)	R_{Th} (Ω)
CALCULATED (Prelab)		
EXPERIMENTAL (Measured)		

(3) Connect the circuit shown in Fig. 8.3. Set the magnitudes of V_{Th} and R_{Th} to the measured values in Table 8.1.

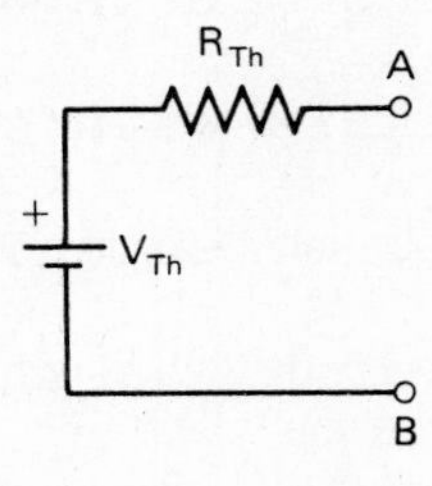

Fig. 8.3

Connect the load resistance (R_L) between the terminals A-B with the necessary instruments to measure I_L and V_L. For each setting of R_L as shown in Table 8.2, measure and record the values of I_L and V_L in the same table.

Table 8.2

	PRELAB CALCULATIONS				EXPERIMENTAL RESULTS			
	Loop-Current Analysis		Thevenin's Circuit		Series-Parallel Circuit		Thevenin's Circuit	
R_L (Ω)	V_L (V)	I_L (mA)	V_L (V)	I_L (mA)	V_L (V)	I_L (mA)	V_L (V)	I_L (mA)
50								
100								
1000								
1250								
1500								
10000								
20000								
30000								

8.5 Comments and Conclusions:

1. Explain 'Thevenin's Theorem' in words.

© 1987 by Prentice-Hall, Inc., A Division of Simon & Schuster, Englewood Cliffs, N.J. 07632. All rights reserved. Printed in the United States of America.

2. Draw the 'Norton's equivalent' circuit for the network in Fig. 8.2.

3. For what value of R_L, is maximum power transferred to the load? Calculate the maximum power transferred in the circuit.

4. Calculate the power dissipated at values of R_L equal to 1 kΩ , 1.25 kΩ and 1.5 kΩ . Are these values less or more than the maximum power? Explain.

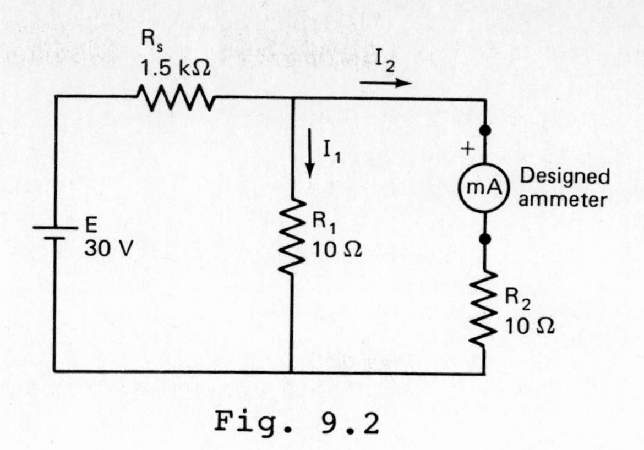

Fig. 9.2

Prelab Work Space:

© 1987 by Prentice-Hall, Inc., A Division of Simon & Schuster, Englewood Cliffs, N.J. 07632. All rights reserved. Printed in the United States of America.

Prelab Work Space:

9.3 Equipment:

ITEM	MANUFACTURER AND MODEL NO.	LAB. SERIAL NO
DC Power Supply		
VOM meter		
DMM or DC milliammeter		
Metermovement		
Three Decade Resistance Boxes		

Resistors: Two 10 Ω , One 4.7 kΩ and 6.8 kΩ .

9.4 Procedure:

A: Characteristics of Metermovement:

(1) Connect the circuit shown in Fig. 9.3 with resistance R_1 set at its maximum value.

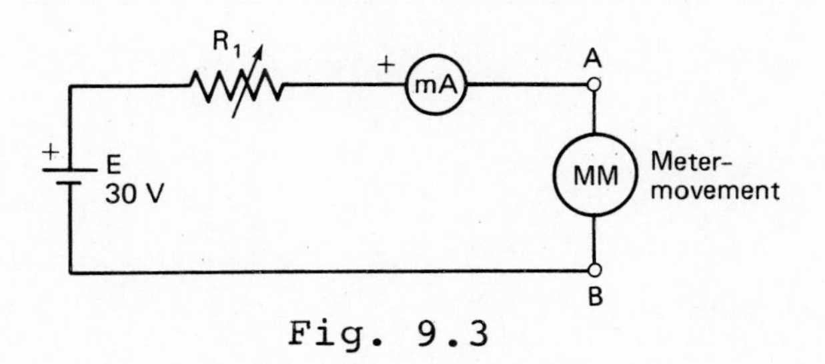

Fig. 9.3

(2) Adjust R_1 until the metermovement (MM) indicates full-scale deflection (fsd). Measure and record the current required for full-scale deflection in Table 9.1.

(3) Connect another decade resistance box (R_2) across the AB terminals in parallel with the

© 1987 by Prentice-Hall, Inc., A Division of Simon & Schuster, Englewood Cliffs, N.J. 07632. All rights reserved. Printed in the United States of America.

metermovement. Start with $R_2 = 0$ and gradually increase R_2 until the MM indicates half full-scale deflection. [The reading on the milliammeter must be maintained at the same value as in Part 2; R_1 may have to be adjusted.] The value of R_2 is now equal to the internal resistance (R_m) of the meter-movement. Record this value in Table 9.1.

Table 9.1

I_{fsd} (mA)	
R_m (Ω)	
Sensitivity of metermovement $S_m = \frac{1}{I_{fsd}} =$ kΩ /V	

B. Voltmeter design and calibration:

(1) Design a voltmeter with a 0-10V range using the MM of Part A. Check your design calculations with the instructor before proceeding. Show all the calculations and the schematic of the voltmeter circuit. Construct the meter.

(2) To calibrate the 'designed voltmeter', use the VOM or DMM as a standard voltmeter. Set up the circuit as shown in Fig. 9.4 and for various values of E, read and record the values of the readings of both meters in Table 9.2.

© 1987 by Prentice-Hall, Inc., A Division of Simon & Schuster, Englewood Cliffs, N.J. 07632. All rights reserved. Printed in the United States of America.

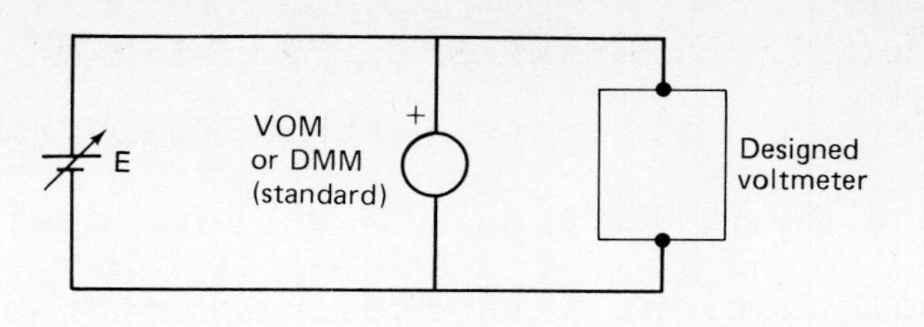

Fig. 9.4

Table 9.2

$V_{Standard}$ (Volts)	$V_{designed}$ (Volts)	% Accuracy
0		
1		
2		
3		
4		
5		
6		
7		
8		
9		
10		

(3) Plot the 'calibration chart' (designed-voltmeter reading versus standard-meter reading) on Graph 9.1.

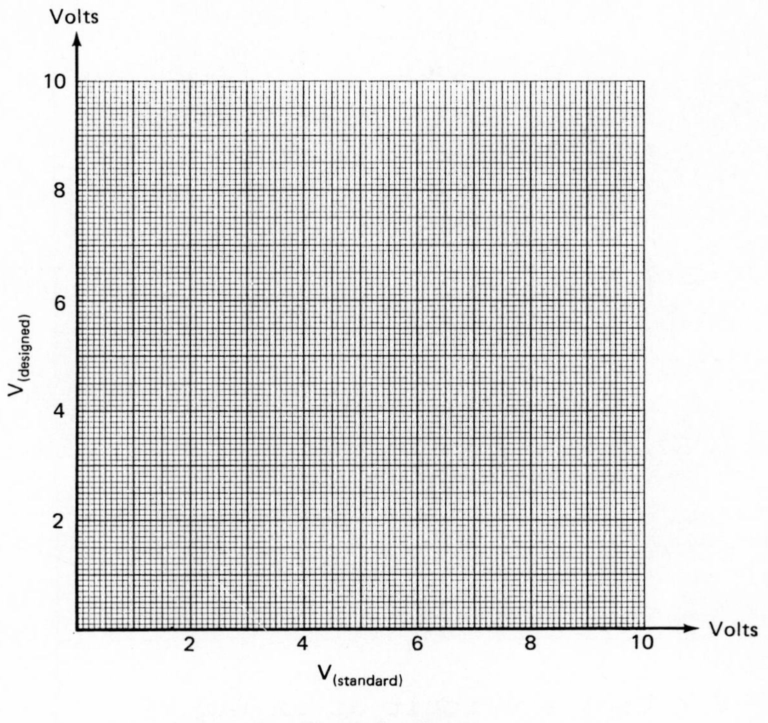

Graph 9.1

(4) Set up the circuit as in Fig. 9.1. Measure the voltage across R_2 using the 'designed-voltmeter'. Calculate the percent accuracy of the meter assuming the theoretical value of V_2 as being the standard value.

% Accuracy =

© 1987 by Prentice-Hall, Inc., A Division of Simon & Schuster, Englewood Cliffs, N.J. 07632. All rights reserved. Printed in the United States of America.

(5) Design and construct a voltmeter with 0-25V range using MM of Part A. Show the schematic of the meter circuit.

(6) Set up the circuit as in Fig. 9.1 and measure the voltage across R_2 using the above 'designed-voltmeter'. Calculate the percent accuracy of this meter.

% Accuracy =

C. Ammeter design and calibration:

(1) Design a milliammeter with a 0-10 mA range using the MM of Part A. Check your design with the instructor before proceeding. Draw the schematic of the ammeter circuit. Construct the meter.

(2) To calibrate the 'designed milliammeter', use the DMM or milliammeter as the 'standard meter'. Set up the circuit as shown in Fig. 9.5. Vary the current supplied from the source (by varying R) in steps of 1 mA. Record the 'standard milliammeter' and 'designed-milliammeter' readings in Table 9.4.

© 1987 by Prentice-Hall, Inc., A Division of Simon & Schuster, Englewood Cliffs, N.J. 07632. All rights reserved. Printed in the United States of America.

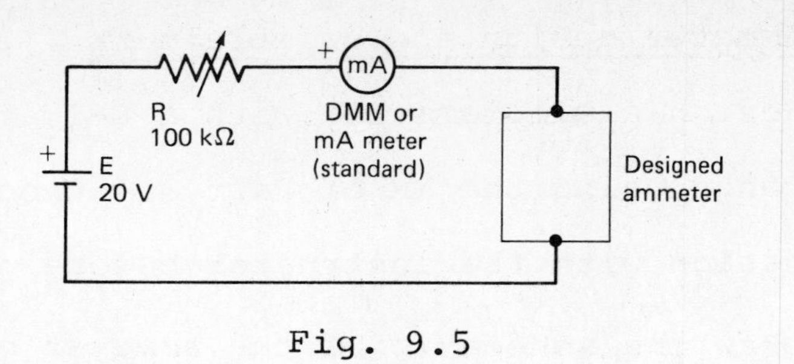

Fig. 9.5

Table 9.4

$I_{Standard}$ (mA)	$I_{designed}$ (mA)	% Accuracy
0		
1		
2		
3		
4		
5		
6		
7		
8		
9		
10		

(3) Plot the 'calibration chart' of the designed meter on Graph 9.2.

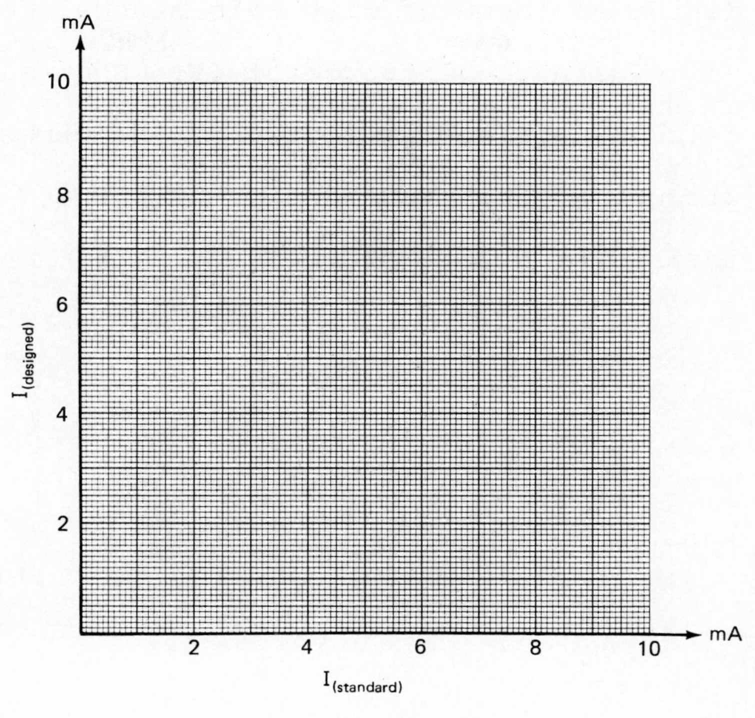

Graph 9.2

(4) Set up the circuit as in Fig. 9.2. Use the 'designed-milliammeter' to measure the current through R_2. Assuming the theoretical value of I_2 as being the standard value, calculate the percent accuracy of the meter.

% Accuracy =

© 1987 by Prentice-Hall, Inc., A Division of Simon & Schuster, Englewood Cliffs, N.J. 07632. All rights reserved. Printed in the United States of America.

D. <u>Design and calibration of an Ohmmeter:</u>

(1) Note the mid-scale ohmic value of your MM; design an ohmmeter with this midscale ohmic value using a DC source of 6V. Show all the calculations and schematic of the ohmmeter circuit. Check your design with the instructor before proceeding.

(2) To calibrate the 'designed-ohmmeter', use a decade resistance box as a standard resistor and set up the circuit as shown in Fig. 9.6.

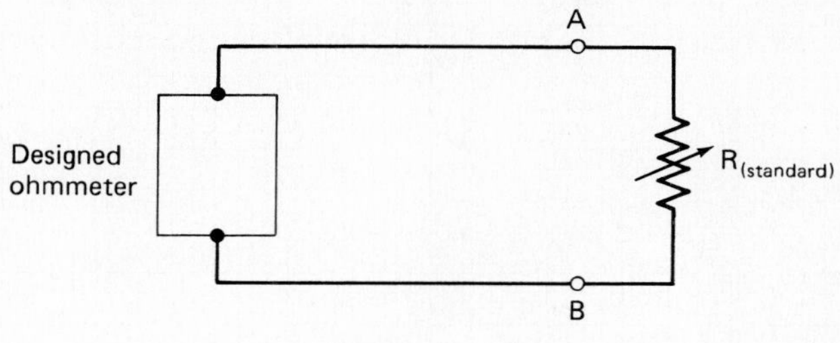

Fig. 9.6

(3) Short-circuit the A-B terminals (R = O) and note the deflection on the 'designed-ohmmeter' (do not adjust any resistances). Remove the short-circuit and vary the values of R as shown in Table 9.5; measure the corresponding values of R using the 'designed- ohmmeter'.

Table 9.5

$R_{standard}$ (Ω)	R (Ω) (designed meter)
0	
50	
200	
500	
800	
1000	
2000	
3000	
5000	
8000	
10000	

© 1987 by Prentice-Hall, Inc., A Division of Simon & Schuster, Englewood Cliffs, N.J. 07632. All rights reserved. Printed in the United States of America.

(4) Plot the 'calibration chart' of the designed ohmmeter on Graph 9.3.

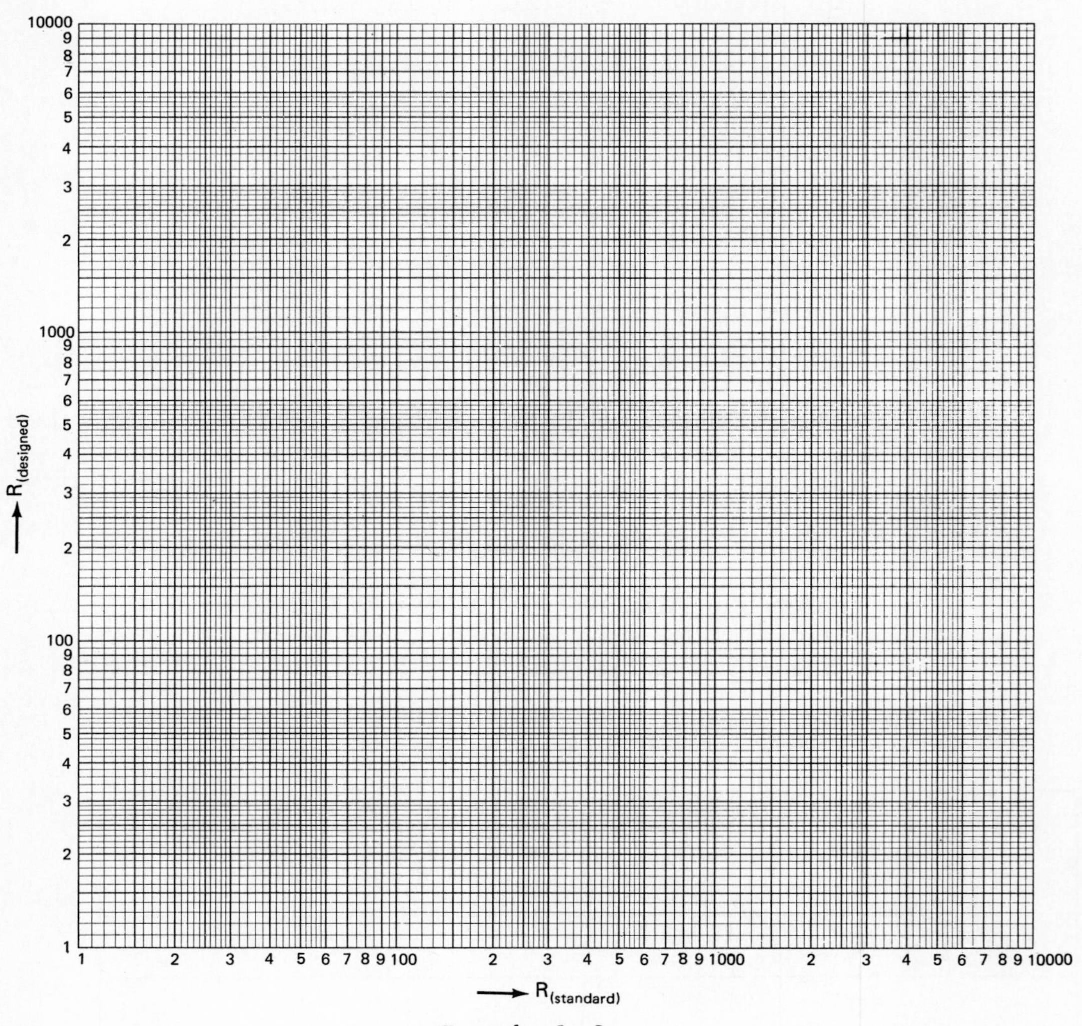

Graph 9.3

9.5 Comments and Conclusions:

1. What is the voltage drop across the MM at the full-scale deflection?

2. Is there a relationship between the sensitivity (S_m) and the internal resistance (R_m) of the metermovement? Explain.

3. What is the sensitivity of the 'designed-voltmeter' as compared to that of the metermovement.

4. What is the sensitivity of the VOM that you used as a standard meter?

© 1987 by Prentice-Hall, Inc., A Division of Simon & Schuster, Englewood Cliffs, N.J. 07632. All rights reserved. Printed in the United States of America.

5. If, in the circuit of Fig. 9.1, R_1 and R_2 are respectively 68 Ω and 47 Ω, instead of 6.8 kΩ and 4.7 kΩ, what is the loading effect of the 'designed-voltmeter' on 10V range?
(Calculate the % accuracy of the meter reading.)

6. If, in the circuit of Fig. 9.1, R_1=680 kΩ and R_2=470 kΩ instead of 6.8 kΩ and 4.7 kΩ, calculate the loading effect of the 'designed-voltmeter' (10V range).

7. What is the sensitivity of the lab-constructed milliammeter as compared with that of the metermovement?

8. A milliammeter with an accuracy of $\pm$ 2% reads 3.5 mA when set on the 10 mA range. Within what range of values (about 3.5 mA) is this meter accurate?

9. The accuracy of any milliammeter is related to the correct value of the shunt resistance (R_{sh}). Consider a nominal 10 mA range milliammeter which consists of a metermovement with the following parameters:

$$S_m = 2.5 \text{ k}\Omega/\text{V}; \; R_m = 500 \; \Omega$$

© 1987 by Prentice-Hall, Inc., A Division of Simon & Schuster, Englewood Cliffs, N.J. 07632. All rights reserved. Printed in the United States of America.

Assume that the value of R_{sh} used, turns out to be 20 Ω (incorrect value). Calculate the percentage error of this milliammeter. Does this milliammeter read too high or too low?

10. Summarize the advantages and disadvantages of measuring an unknown resistance by the three different methods you have used in the lab. (Volt-amp method, Wheatstone bridge and Ohmmeter.)

11. The Ohmmeter readings are more accurate around the mid-scale range and less accurate at the extreme ends of the scale. Explain why.

© 1987 by Prentice-Hall, Inc., A Division of Simon & Schuster, Englewood Cliffs, N.J. 07632. All rights reserved. Printed in the United States of America.

12. What is the purpose of the 'zero-adjust' potentiometer on the VOM (as an ohmmeter)?

13. Why is the 'ohm' scale non-linear and counter-clockwise?

Experiment 10

THE OSCILLOSCOPE AS A CURRENT AND VOLTAGE MEASURING INSTRUMENT

Required Reading: Text, section 7.5

10.1 Objective:

- To gain familiarity with the basic operating controls of a dual-beam oscilloscope.
- To use the oscilloscope for measuring voltages and currents.

10.2 Prelab Assignment:

Read the operating instructions for the dual-beam oscilloscope available at your laboratory. Appendix 10.1 shows, as an example, the operating instructions for the Philips PM-3233 oscilloscope.

© 1987 by Prentice-Hall, Inc., A Division of Simon & Schuster, Englewood Cliffs, N.J. 07632. All rights reserved. Printed in the United States of America.

10.3 Equipment:

ITEM	MANUFACTURER AND MODEL NO.	LAB. SERIAL NO
Dual-Beam Oscilloscope		
Signal Generator		
DC Power Supply		
DMM or VOM		
Decade Resistance Box		

Resistors: One 10Ω

10.4 Procedure:

A. Calibration of the Oscilloscope:

(1) Turn on the oscilloscope after you have become familiar with the locations and functions of its controls.

(2) Set (Time/div) to 0.1 m sec/div and (trigger mode) to (Auto). Adjust the focus and intensity controls to achieve well-defined time-base line displays (one for each beam).

(3) Adjust the vertical-position control to center each beam on the screen; set the (AC - 0 - DC) controls to (AC).

(4) Connect the Cal-terminal of the oscilloscope to the input of channel Y_A. [The Cal-terminal output provides a square waveform with a well-defined (peak-to-peak) voltage.] Adjust the vertical sensitivity of Y_A (V/div) to achieve maximum vertical deflection of the display. Measure the (peak-to-peak) deflection and determine V (p-p).

(5) Repeat Step # 4 for channel Y_B and record your results in Table 10.1.

Table 10.1

Channel #	(p-p) deflection (div)	Sensitivity (V/div)	V (p-p) (V)
Y_A			
Y_B			

How do the V (p-p) values compare with V (p-p) of the (Cal) output?

B. Voltage Measurement:

(1) Use the VOM (or DMM) to adjust the terminal voltage (V_T) of the DC power supply to 1V.

© 1987 by Prentice-Hall, Inc., A Division of Simon & Schuster, Englewood Cliffs, N.J. 07632. All rights reserved. Printed in the United States of America.

(2) Set the (AC - O - DC) control of Y_A to (DC). Connect Y_A to display the terminal voltage of the DC power supply [adjust (V/div) to achieve the maximum vertical deflection]; determine the DC-voltage and record in Table 10.2.

(3) Repeat the above steps for each value of V_T in Table 10.2.

Table 10.2

V_T (V)	1	2	5	10
Vertical Deflection (div)				
Vertical Sensitivity (V/div)				
Measured Voltage by the oscilloscope V_{osc} (V)				

(4) Plot the measured voltage by the oscilloscope V_{osc} versus V_T on Graph 10.1.

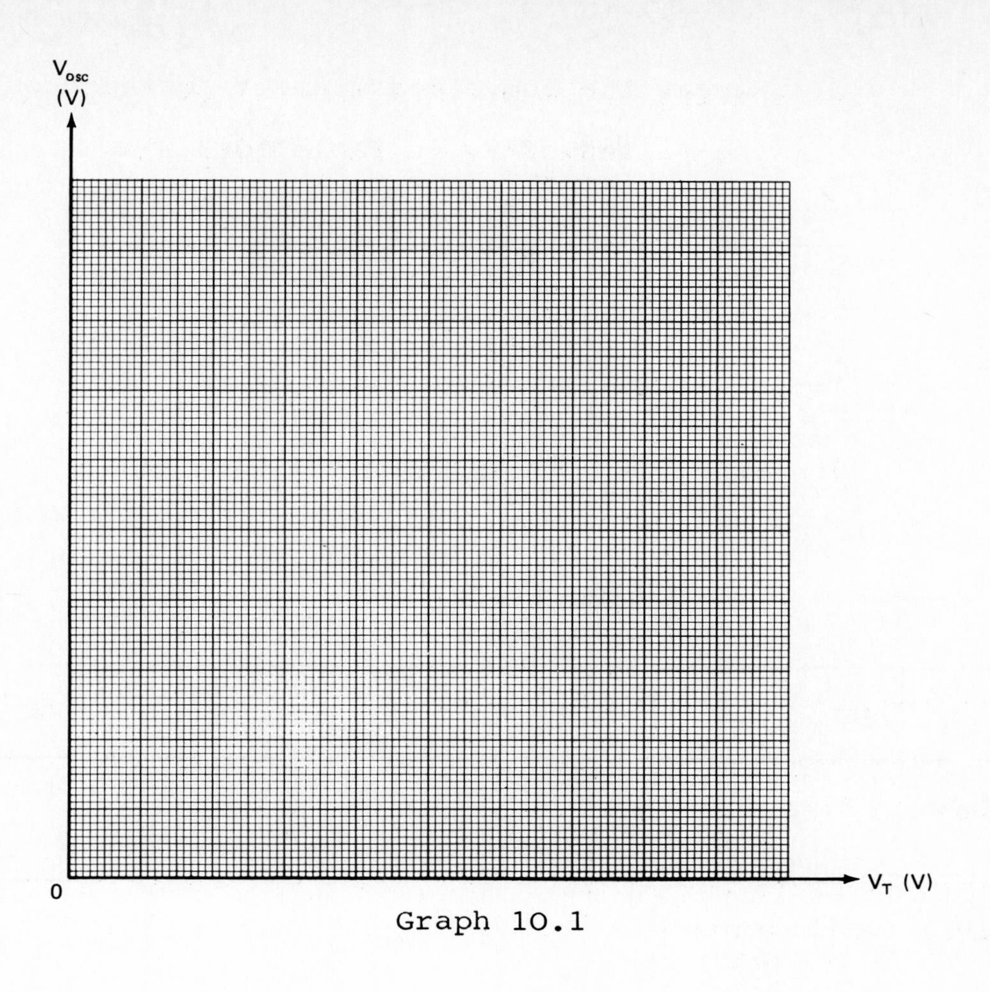

Graph 10.1

C. Current Measurement:

(1) Connect the circuit shown in Fig. 10.1. Set V_T to 10V and adjust the decade resistance box (R_X) to 10 kΩ . Connect Y_A to display the voltage across the 10 Ω -current-sampling resistance [Y_A now displays 10 X I]. Adjust the vertical sensitivity of Y_A to achieve the maximum vertical deflection; determine the DC-current and record your results in Table 10.3.

© 1987 by Prentice-Hall, Inc., A Division of Simon & Schuster, Englewood Cliffs, N.J. 07632. All rights reserved. Printed in the United States of America.

(2) Repeat the above measurement for each of the values of R_X in Table 10.3.

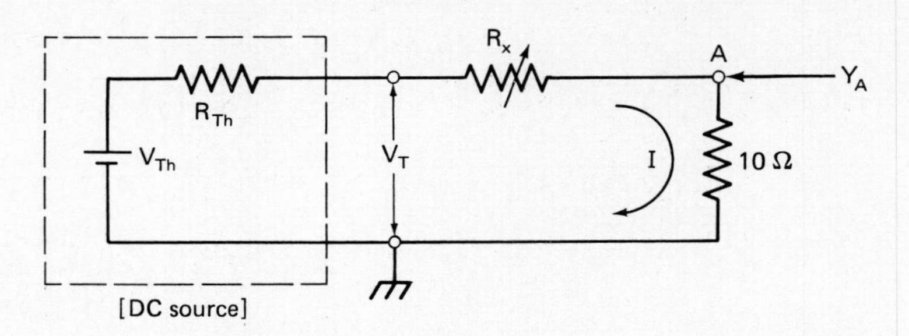

Fig. 10.1

Table 10.3

R_X (k Ω)	10	5	2	1
I (mA)				
Vertical Deflection (div)				
Vertical Sensitivity (V/div)				
Measured Current by the oscilloscope I_{osc} (mA)				

(3) Plot the measured current by the oscilloscope I_{osc} versus I on Graph 10.2.

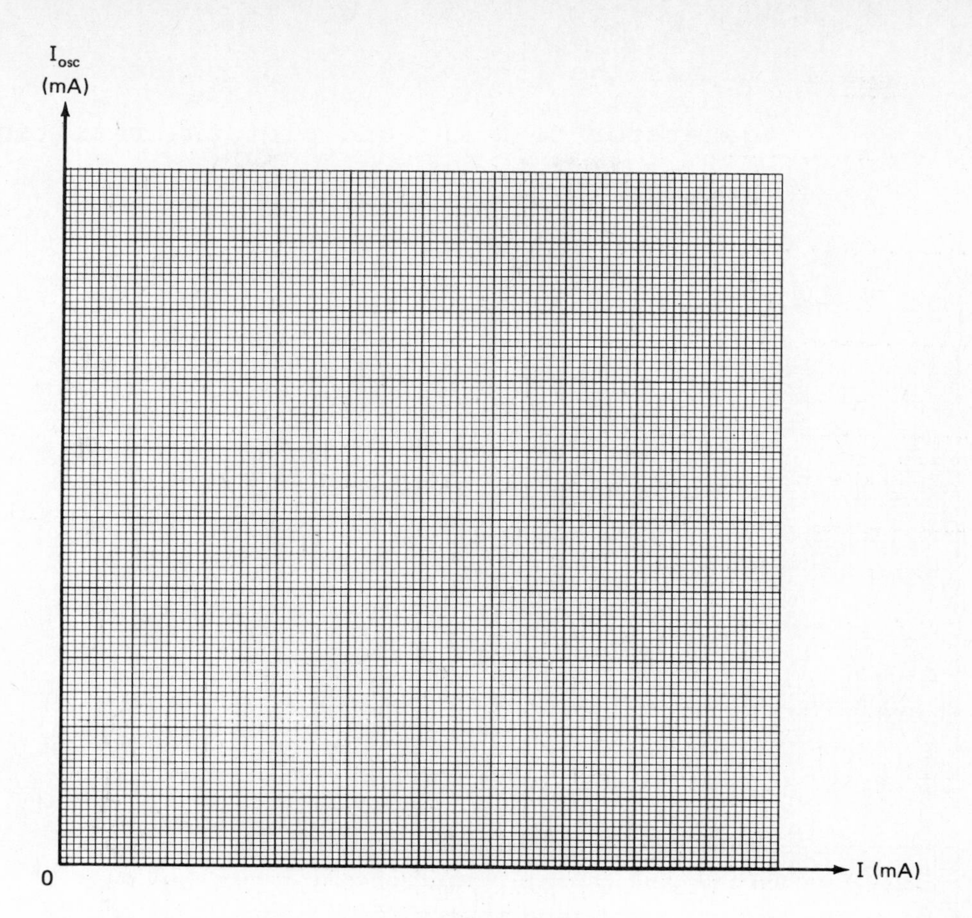

Graph 10.2

D. Signal Display:

(1) Adjust the frequency of the (sinusoidal) signal generator to 1 kHz. Set the vertical sensitivity of Y_A to 1V/div and the (Time/div) control to 0.1 msec/div. Connect Y_A to display the terminal voltage V_T of the sinusoidal generator; adjust V_T to achieve a (peak-to-peak) display of 8V.

(2) Plot V_T versus time on Graph 10.3.

© 1987 by Prentice-Hall, Inc., A Division of Simon & Schuster, Englewood Cliffs, N.J. 07632. All rights reserved. Printed in the United States of America.

(3) Adjust the frequency of the sinusoidal generator to 2 kHz and plot the resulting display on Graph 10.3.

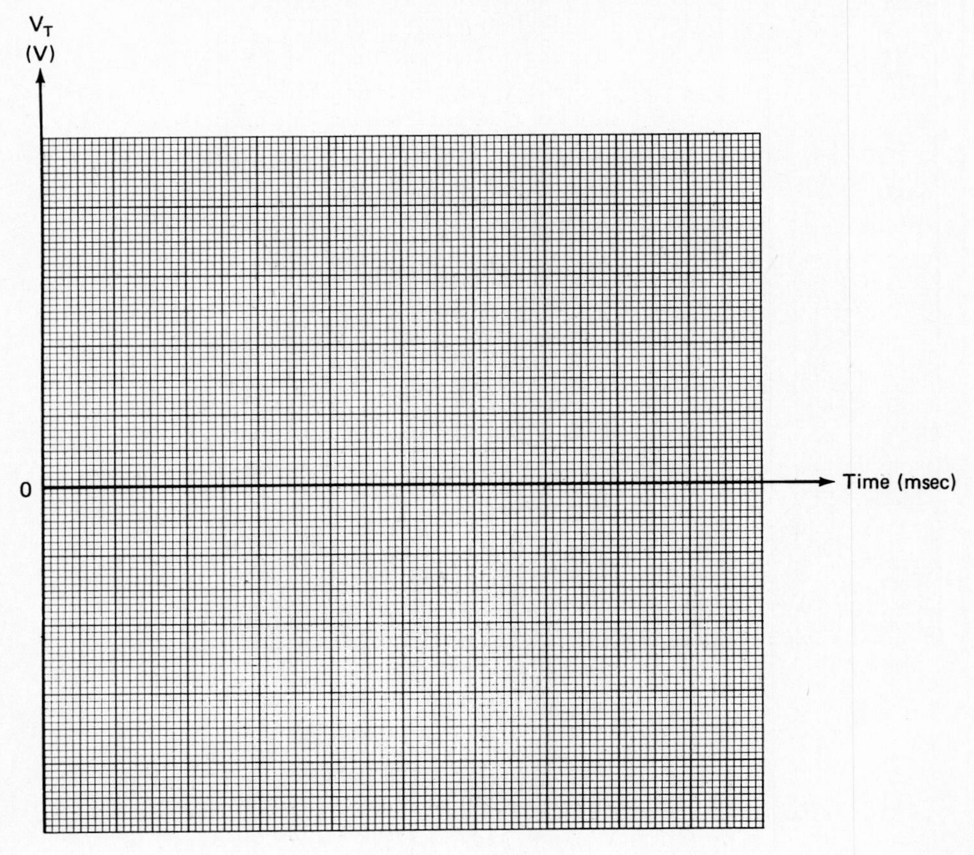

Graph 10.3

10.5 Comments and Conclusions:

1. What is the maximum possible voltage that can be measured with your oscilloscope?

2. Assume that a minimum deflection of 0.1 of a division can be estimated accurately. What is the minimum voltage that can be accurately measured by your oscilloscope?

3. Consider the circuit shown in Fig. 10.2. Assume that the input resistance of Y_A is 1 $M\Omega$ and the vertical sensitivity is set at 0.2V/div. Determine:

a. the number of vertical-divisions of the corresponding beam deflection.

b. the voltage measured by the oscilloscope.

c. the voltage measured by the oscilloscope, assuming that the input resistance of Y_A is increased to 10 $M\Omega$.

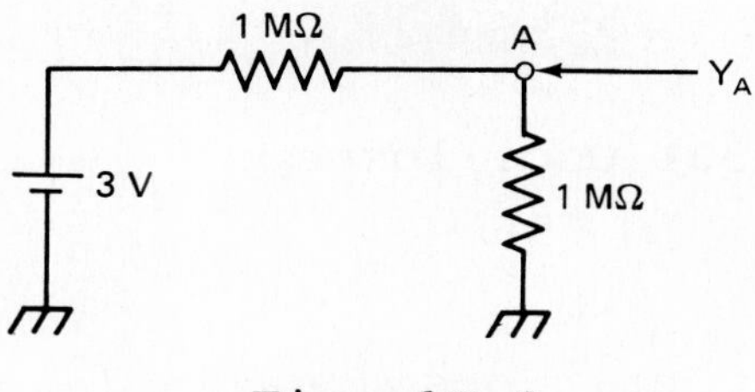

Fig. 10.2

© 1987 by Prentice-Hall, Inc., A Division of Simon & Schuster, Englewood Cliffs, N.J. 07632. All rights reserved. Printed in the United States of America.

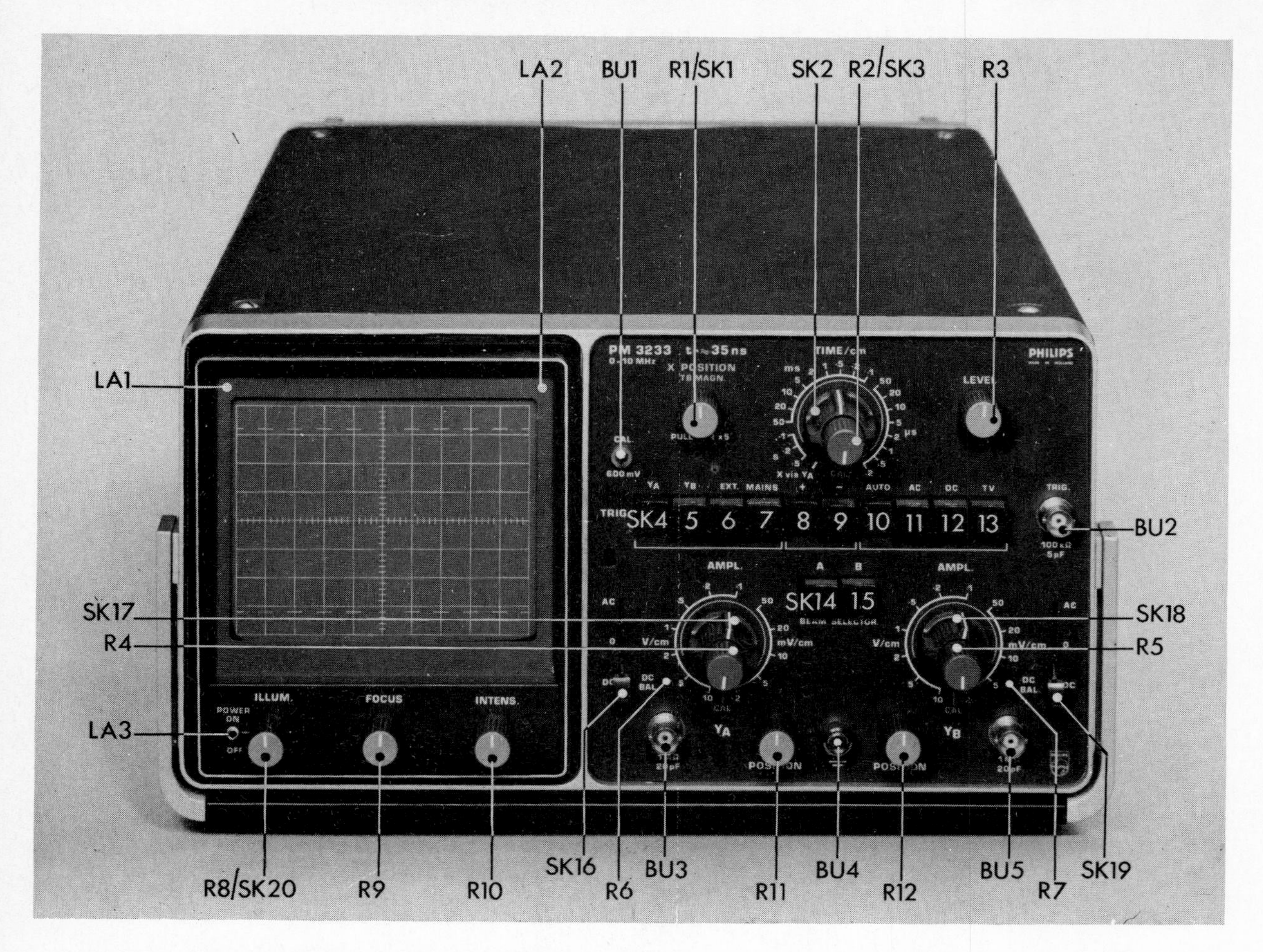

Philips (PM-3233) Oscilloscope *Courtesy, Philips Electronics Ltd.*

CONTROLS AND SOCKETS

X POSITION (R1) — Continuously variable control giving horizontal positioning of the display.

MAGN' (SK1) — Incorporates a switch for calibrated 5x magnification of the time base.

TIME/cm (SK2) — Time-coefficient control of the time-base; 21 way switch with a position for external X deflection (X via Y_A).

CAL.-TIME/cm (R2/SK3)
Continuously variable control of the time coefficients.
In the CAL. position the time-coefficient is calibrated.

LEVEL (R3) — Continuously variable control to select the level at which the time-base generator starts.

CAL. (BU1) — Outlet for square-wave voltage of 600 mV p-p for calibration purposes.

TRIGGERING (SK4..13) — Controls for trigger source, slope and mode; 10-way push-button switch.

Y_A (SK4) — Internal triggering signal derived from channel Y_A.

Y_B (SK5) — Internal triggering signal derived from channel Y_B.

EXT. (SK6) — Triggering signal derived from a voltage applied to the TRIGG. input socket.

MAINS (SK7) — Triggering signal derived from an internal voltage with mains frequency. This trigger source is inoperative when the instrument is supplied with an external DC voltage.

+ (SK8) — Provides for triggering on the positive slope of the signal.

- (SK9) — Provides for triggering on the negative slope of the signal.

AUTO (SK10) — Provides for a free-running time-base in the absence of triggering signals and automatic signal-derived limitation on the LEVEL control range.

AC (SK11) — Triggering with coupling capacitor in triggering-signal path.

DC (SK12) — Direct coupling for triggering on a slowly varying voltage or for full-bandwidth working.

TV (SK13) — Enables triggering on either line or frame pulses of TV signals, as dictated by the position of TIME/cm switch SK2. Triggering on frame pulses in positions 50 μs/cm to 0.5 s/cm and on line pulses in position 0.2 μs/cm to 20 μs/cm.

TRIGG. (BU2) — Input BNC socket for external triggering signals.

BEAM SELECTOR A (SK 14) — If this push-button is depressed, vertical deflection is achieved by the signal connected to the channel Y_A input.

© 1987 by Prentice-Hall, Inc., A Division of Simon & Schuster, Englewood Cliffs, N.J. 07632. All rights reserved. Printed in the United States of America.

BEAM SELECTOR A (SK 15)

If this push-button is depressed, vertical deflection is achieved by the signal connected to the channel Y_B input.

If both switch A (SK 14) and switch B (SK 15) are depressed, vertical deflection is achieved by both the signal connected to the channel Y_A input and the signal connected to the channel Y_B input.

AC-O-DC (SK 16 & 19)

Signal coupling, three position switch.
AC : via coupling capacitor
O : interruption of connection between input socket and input circuit, the latter being earthed.
DC : Direct coupling

AMPL. (SK 17 & 18) Control of the vertical deflection coefficients, 12-way switch.

CAL.-AMPL. (R4 & 5) Continuously variable control of the vertical deflection coefficients. In the CAL. position, the deflection coefficient is calibrated.

DC BAL. (R6 & 7) Continuously variable control of the direct-voltage balance of the vertical amplifier.

ILLUM. (SK 20 & R8) Continuously variable control of the graticule illumination.
Incorporates mains switch.

FOCUS (R9) Continuously variable control of the electron-beam focusing.

INTENS (R10) Continuously variable control of the trace brilliance.

1 MOhm-20 pF (BU3 & 5)

Input BNC socket for the vertical deflection signals.

POSITION (R11 & 12)

Continuously variable control giving vertical positioning of the display.

(BU4) Earth socket.

TRIGGERING

General

In order to obtain a stationary trace, the horizontal deflection must always be started at a fixed point of the signal. The time-base generator is, therefore started by narrow trigger pulses formed in the trigger pulse shaper (Schmitt trigger), controlled by a signal originating from the vertical input signal or an external source.

Trigger coupling

AC If the signal voltage contains a DC component triggering can cease when the level potentiometer cannot supply the correct DC level for the Schmitt trigger. In this case it is useful to apply AC coupling. AC coupling is obtained by inserting a capacitor in the trigger path. This means that the signal can still be DC coupled to the Y channels.

DC DC Coupling is useful when the mean value of this signal varies. This sort of signal often occurs in digital systems. With AC coupling the trigger point would not be fixed which would give rise to jitter or even loss of triggering.

Trigger level

In case of a complicated signal in which a number of non-identical voltage shapes occur periodically, the time axis should always be started with the same voltage shape so as to obtain a stationary trace. This is possible when one of the details has a deviating amplitude. By means of the LEVEL knob, the trigger level can be set in such a way that only this larger voltage variation passes this level. The LEVEL control is also very useful when two signals must be compared accurately e.g. in phase measurements. By means of the LEVEL control the starting point of the traces can then be shifted exactly on to the central graticule line.

Automatic triggering

Automatic triggering (when the AUTO switch is depressed) is most often used on account of its simple operation.
In this mode it is possible to display a large variety of waveforms having different amplitude and shape, without it being necessary to operate any of the trigger controls.
If no triggering signal is present, a time-base line remains visible on the screen. This is useful for zero reference purposes. In this trigger mode the level can be adjusted over the peak-to-peak value of the AC component of the

© 1987 by Prentice-Hall, Inc., A Division of Simon & Schuster, Englewood Cliffs, N.J. 07632. All rights reserved. Printed in the United States of America.

signal. If none of the switches AUTO, AC, DC or TV is depressed, the oscilloscope works in the automatic mode, but with the entire level range available. This has the advantage that there is always a trace visible, even when no TRIGG. push-buttons are depressed.

External triggering

External triggering is applied for signals having a strongly varying amplitude, if a signal having a fixed amplitude and equal frequency is available. Even more important is external triggering in case of complex signals and pulse patterns. Then external triggering can be used to avoid double traces.
This obviates the necessity of readjusting the level setting at every variation of the input signal.

Triggering with the mains frequency

In this case the triggering signal is a sine-wave with the mains frequency. This trigger source is useful if the frequency of the signal under observation is coupled with the mains frequency.
It is, e.g., possible to recognize the hum component of a signal by triggering on that component.

Triggering with television signals

It is possible to trigger on the line or frame sync pulses of television signals. In positions .5 s to 50 μs of the TIME/cm switch triggering takes place on the frame sync pulses and in positions 20 μs to .2 μs on the line sync pulses of the signal.
The position of the trigger slope switches must correspond to the polarity of the video information of the signal.

TIME-BASE MAGNIFIER

The magnifier is operated with a push-pull switch.
When this switch is in the x5 position, the time-base sweep speed is increased 5 times. In this position the sweep time is determined by dividing the indicated TIME/cm value by 5.

Z MODULATION

In order to bring extra information in the C.R.T. display without changing the form of the display, the brightness of the trace can be lowered by an externally applied voltage.
The external signal must, therefore, be fed to the Z MOD socket at the rear of the oscilloscope.
The voltage required for visible brightness modulation depends on the position of the INTENS control.
With an average brightness of the trace, a 20 V_{p-p} voltage is amply sufficient for obtaining a good visible "Z-modulation".

THE DUAL-BEAM TUBE

The cathode-ray tube used in PM 3233 oscilloscope is a dual-beam tube in which two beams are generated in one gun and can be deflected independently.
This arrangement is also known as split-beam tube.

In this tube, both time-base lines are exactly in parallel as they originate from one point and are under the influence of one common horizontal amplifier. Because the two traces originate from one gun, they show little distortion in relation to one another.

The split-beam tube is very suitable for displaying signals with a low repetition rate at relatively fast sweep speeds, since it may be regarded as a tube with a chopper and an infinitely high chopper frequency.

For equalizing and adjusting the brightness of both beams, two magnets are mounted symmetrically on the C.R.T. One of these magnets can be readjusted by means of a screwdriver through an opening in the bottom-plate of the oscilloscope.

. BRIEF CHECKING PROCEDURE

STARTING POSITIONS OF THE CONTROLS

- Push-buttons Y_A SK4, + SK8 and BEAM SELECTOR A SK14 & B SK 15 depressed.
- TIME/cm switch SK2 to .1 ms
- AMPL switches SK 17 & SK 18 to .1 V/cm
- MAGN switch SK 1 to x1
- POSITION potentiometers R1, R11 and R12 to their mid-positions
- INTENS potentiometer R1O fully clockwise
- TIME/cm and AMPL potentiometers R2, R4 and R5 to CAL

Unless otherwise stated, the controls always occupy the same position as in the previous check.

© 1987 by Prentice-Hall, Inc., A Division of Simon & Schuster, Englewood Cliffs, N.J. 07632. All rights reserved. Printed in the United States of America.

Experiment 11

THE OSCILLOSCOPE FOR SIGNALS DISPLAY

Required Reading: Text, section 7.5

11.1 Objective:

- To gain familiarity with the triggering and synchronization controls of the oscilloscope.
- To examine the effects of circuit ground on signals display.

11.2 Prelab Assignment:

(1) A sinusoidal signal of 400 Hz is to be displayed on the oscilloscope, using channel Y_A, such that:

- two complete cycles of the signal are to appear on the screen
- the starting-point of the display is a positive-slope zero-crossing point

© 1987 by Prentice-Hall, Inc., A Division of Simon & Schuster, Englewood Cliffs, N.J. 07632. All rights reserved. Printed in the United States of America.

What are the proper settings for:

- the (trigger-source) control,
- the (Time/div) control, and
- the (level) control?

(2) Consider the circuit shown in Fig. 11.1. A dual-beam oscilloscope is used to display the voltage differences between the various nodes of the circuit. Initially the time-base lines for both beams are set at the center of the screen; the vertical sensitivities for Y_A and Y_B are set at 1V/div. The (Time/div) control is set at 1 msec/div.

Plot the resulting displays of both beams, for each of the following conditions.

Condition #	Y_A at node #	Y_B at node #	Ground at node #
1	x	y	z
2	x	z	y
3	z	y	x

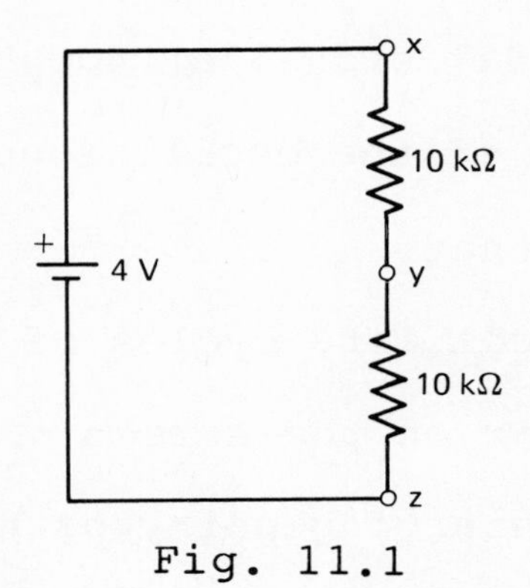

Fig. 11.1

Prelab Work Space:

© 1987 by Prentice-Hall, Inc., A Division of Simon & Schuster, Englewood Cliffs, N.J. 07632. All rights reserved. Printed in the United States of America.

Prelab Work Space:

11.3 Equipment:

ITEM	MANUFACTURER AND MODEL NO.	LAB. SERIAL NO
Dual-Beam Oscilloscope		
Signal Generator		
DC Power Supply		
DMM or VOM		
Decade Resistance Box		

Resistors: Two 10 kΩ

11.4 Procedure:

A. Signal Display and Synchronization:

(1) Set the controls of the oscilloscope to the following positions:

- Triggering-source....... Y_B
  - -mode .... Auto
  - -slope .... (+)
- Time/div .... 0.5 msec/div
- For both beams:
  - vertical sensitivity ... 1V/div
  - (AC-O-DC) control ...... (O)
  - Y-position .... center of screen

(2) Connect the time-base output terminal [usually located at the back panel of the oscilloscope] to Y_A. Set the (AC-O-DC) control of Y_A to (DC). The A-beam now displays one cycle of the sawtooth waveform from the internal sweep generator.

(3) Adjust the (X-position) control to move the starting point of the display to the extreme left-hand vertical graticule line. Plot the display on Graph 11.1.

© 1987 by Prentice-Hall, Inc., A Division of Simon & Schuster, Englewood Cliffs, N.J. 07632. All rights reserved. Printed in the United States of America.

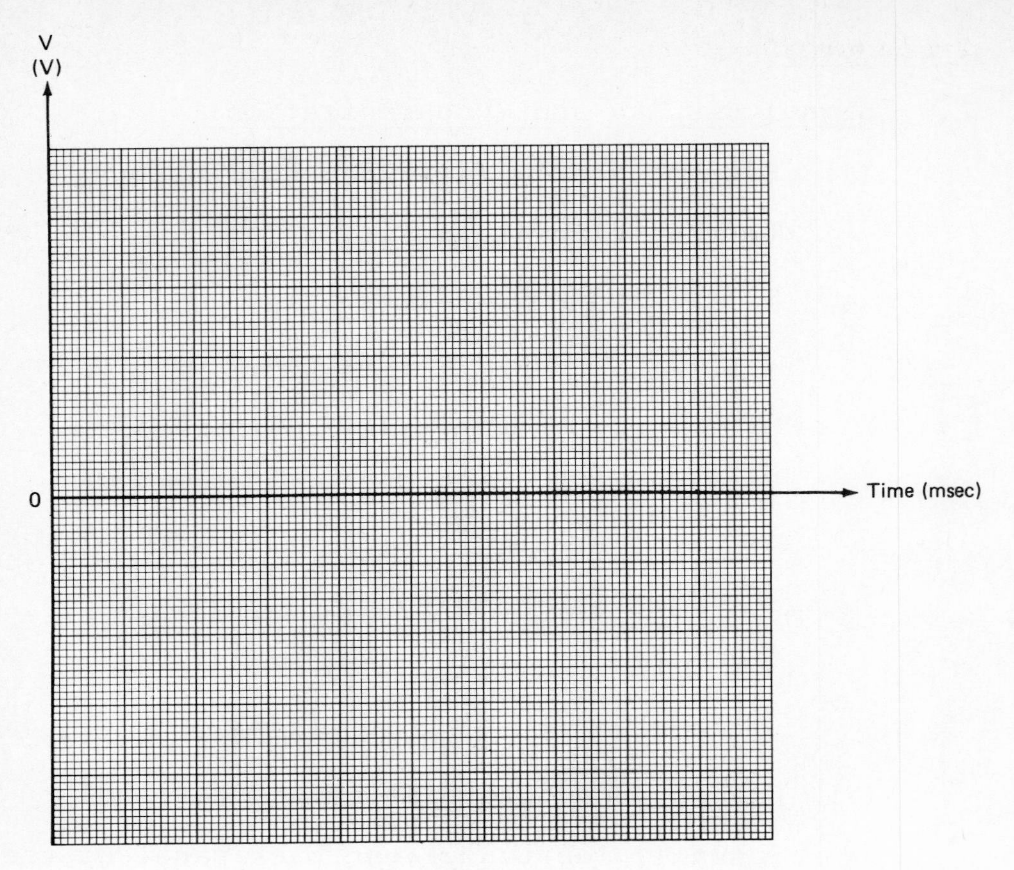

Graph 11.1

(4) Adjust the frequency of the sinusoidal-signal generator to 400 Hz; connect Y_B to display the terminal voltage V_T of the generator. Set the (AC-O-DC) control of Y_B to (AC) and adjust V_T to achieve a peak-to-peak display of 6V.

(5) Vary the position of the (level) control and observe its effect on the displays of both beams. Comment on your results.

(6) Adjust the (level) control until the starting (extreme left-hand) point of the Y_B display is a positive-slope zero-crossing point. Plot the display of Y_B on Graph 11.1.

(7) Set the triggering slope to (-); how does this change affect the displays?

(8) Set the triggering mode to free running and vary the (level) control. How does this change affect the displays?

(9) Set the triggering mode and slope back to (Auto) and (+), respectively. Adjust the (Time/div) control to 1 msec/div. How does this change affect the displays and how many trigger pulses does the sweep generator ignore during each sweep cycle?

© 1987 by Prentice-Hall, Inc., A Division of Simon & Schuster, Englewood Cliffs, N.J. 07632. All rights reserved. Printed in the United States of America.

B. <u>Effects of Circuit Ground:</u>

(1) Connect the circuit shown in Fig. 11.2. Use the VOM (or DMM) to adjust V_T to 4V.

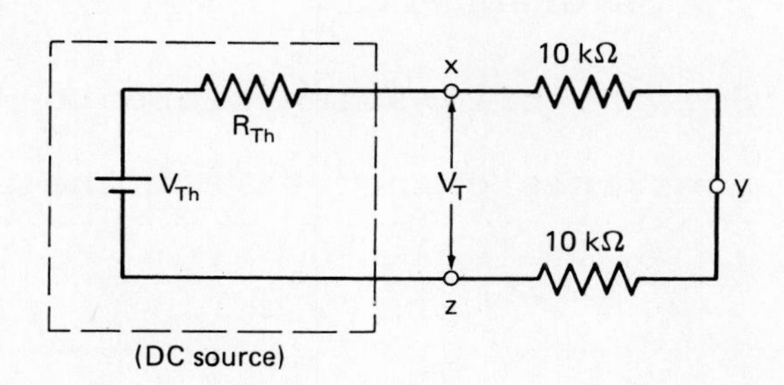

Fig. 11.2

(2) Set both (AC-O-DC) controls to (DC). Connect Y_A to node x, Y_B to node y and the oscilloscope ground terminal to node z. Measure V_x and V_y w.r.t. node z and record your results in Table 11.1.

(3) Repeat Step # 2 for each of the conditions indicated in Table 11.1.

Table 11.1

Y_A @ node #	Y_B @ node #	ground @ node #	V_x (V)	V_y (V)	V_z (V)	w.r.t. node #
x	y	z				z
x	z	y				y
z	y	x				x

(4) Connect the circuit shown in Fig. 11.3; Set V_T to 25 V. The circuit has a fixed ground connection that is common with the oscilloscope ground. Use your oscilloscope [WITH THE GROUND-TERMINAL OF THE OSCILLOSCOPE PERMANENTLY CONNECTED TO THE CIRCUIT GROUND] to determine the voltages across the source terminals V_T, R_1, R_2 and R_3. Record your results in Table 11.2.

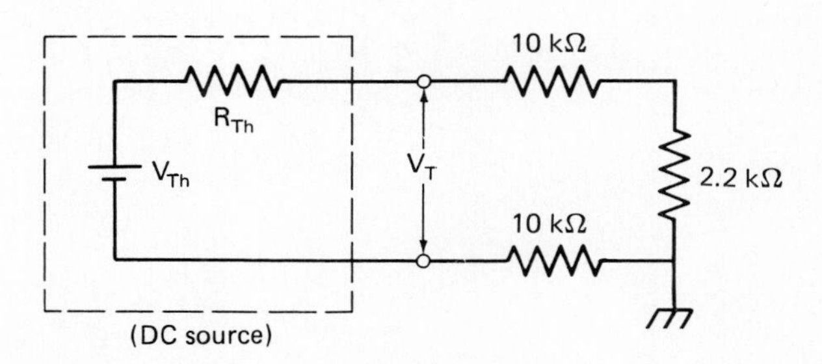

Fig. 11.3

© 1987 by Prentice-Hall, Inc., A Division of Simon & Schuster, Englewood Cliffs, N.J. 07632. All rights reserved. Printed in the United States of America.

Table 11.2

V_T (V)	V_{R_1} (V)	V_{R_2} (V)	V_{R_3} (V)

Experiment 12

THE OSCILLOSCOPE AS A CURVE TRACER

12.1 Objective:

- To examine the XY-mode operation of the oscilloscope.
- To display the I-V characteristics of two-terminal devices.

12.2 (a) Introduction:

Experiment #1 dealt with the I-V characteristics of a resistor. The characteristics were obtained by taking a number of measurements of the voltage across and the current through the resistor, at different settings of the source voltage. This procedure is rather tedious and time consuming.

A quick and a more accurate method of obtaining the I-V characteristics involves the use of an oscilloscope together with a sweep voltage source as shown in Fig. 12.1.

© 1987 by Prentice-Hall, Inc., A Division of Simon & Schuster, Englewood Cliffs, N.J. 07632. All rights reserved. Printed in the United States of America.

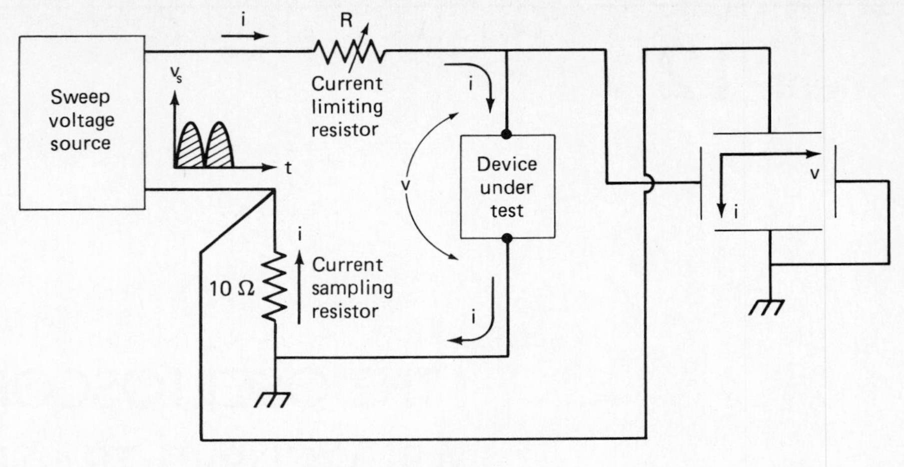

Fig. 12.1

The voltage across the current-sampling resistor is applied to the vertical-deflecting plates of the scope; hence, the vertical position of the electron beam is directly proportional to "i." Similarly, the voltage across the device is applied to the horizontal-deflecting plates; therefore, the horizontal position of the beam is directly proportional to "v." Thus, v & i control the x and y coordinates of the electron beam. A change in the value of the sweep voltage, v_s, will lead to a new set of coordinates for the I-V characteristics, displayed on the screen of the oscilloscope.

The sweep voltage must satisfy three basic requirements here:

(a) it must be repetitive to ensure retracing of the I-V graph.

(b) its repetition rate must be sufficiently fast to avoid any flickering effect.

(c) it must have a continuous amplitude range (square wave is no good).

The current-limiting resistor R, is there to ensure the safety of the device under test.

12.2 (b) **Prelab Assignment:**

Given the I-V characteristics of a two-terminal device as shown in Fig. 12.2:

(1) determine graphically (using the load-line approach) the coordinates (V & I) of the operating point for the following settings of the source voltage:

$$0, \pm 1, \pm 2V.$$

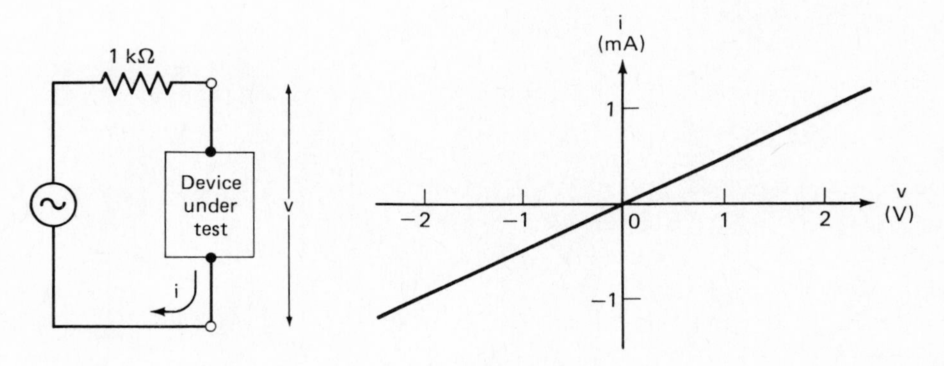

Fig. 12.2

© 1987 by Prentice-Hall, Inc., A Division of Simon & Schuster, Englewood Cliffs, N.J. 07632. All rights reserved. Printed in the United States of America.

Prelab Work Space:

12.3 Equipment:

ITEM	MANUFACTURER AND MODEL NO.	LAB. SERIAL NO
Dual-Beam Oscilloscope		
Signal Generator		

Resistors: One 1 kΩ & 10 kΩ

Diodes: IN4001 (or equivalent)

12.4 Procedure:

A. Setting the Oscilloscope in the XY-Mode:

(1) Set the appropriate control of your oscilloscope to allow access to the input terminals of the X-amplifier. For example, setting the (Time/div) control of the Philips PM3233 Oscilloscope at "X via Y_A" converts Y_A to be the input terminals of the X-amplifier (X-input), i.e., any signal applied now to Y_A will control the X-position of beam B [beam A does not exist in the XY-mode].

(2) Set the controls for Y_B as follows:

. vertical sensitivity .... 2 mV/div

. (AC-O-DC) .... (O)

Set the controls for X-input at:

. sensitivity................. 0.5V/div

. (AC-O-DC)................... (O)

© 1987 by Prentice-Hall, Inc., A Division of Simon & Schuster, Englewood Cliffs, N.J. 07632. All rights reserved. Printed in the United States of America.

(3) Adjust the (Y_B-position) and the (X-position) controls to place the beam at the center of the screen; the beam YX-coordinates are now considered as (0, 0).

B. <u>Obtaining the I-V Characteristics for Two-Terminal Devices:</u>

(1) Adjust the frequency of the sinusoidal generator to about 200 Hz. Use the oscilloscope to adjust the terminal voltage of the generator (V_T) to 4V (peak-to-peak).

(2) Connect the circuit shown in Fig. 12.3; the two-terminal device under test is a 1 kΩ resistor.

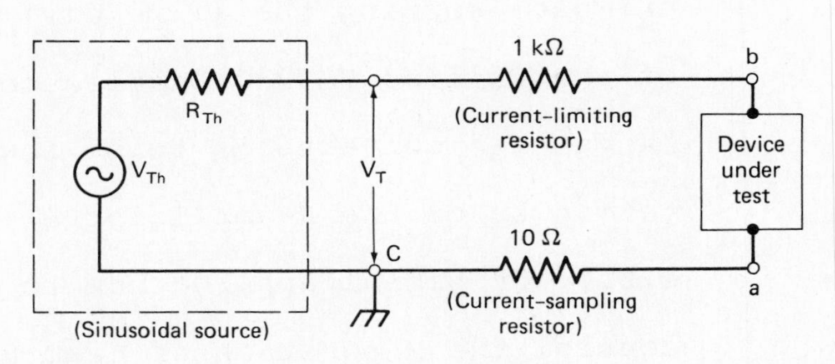

Fig. 12.3

(3) Connect node (a) to the Y_B-input, node (b) to the X-input and node (c) to the oscilloscope ground terminal. Set both (AC-O-DC) controls to (DC). The X and Y deflections of the beam (w.r.t. the center of the screen) represent the voltage across and 10X the current

through the device, respectively.

(4) Plot the I-V characteristics of the device under test on Graph 12.1.

(5) Repeat the above steps for each of the following devices:

- a 10 kΩ resistor
- a semiconductor diode (IN4001 or equivalent).

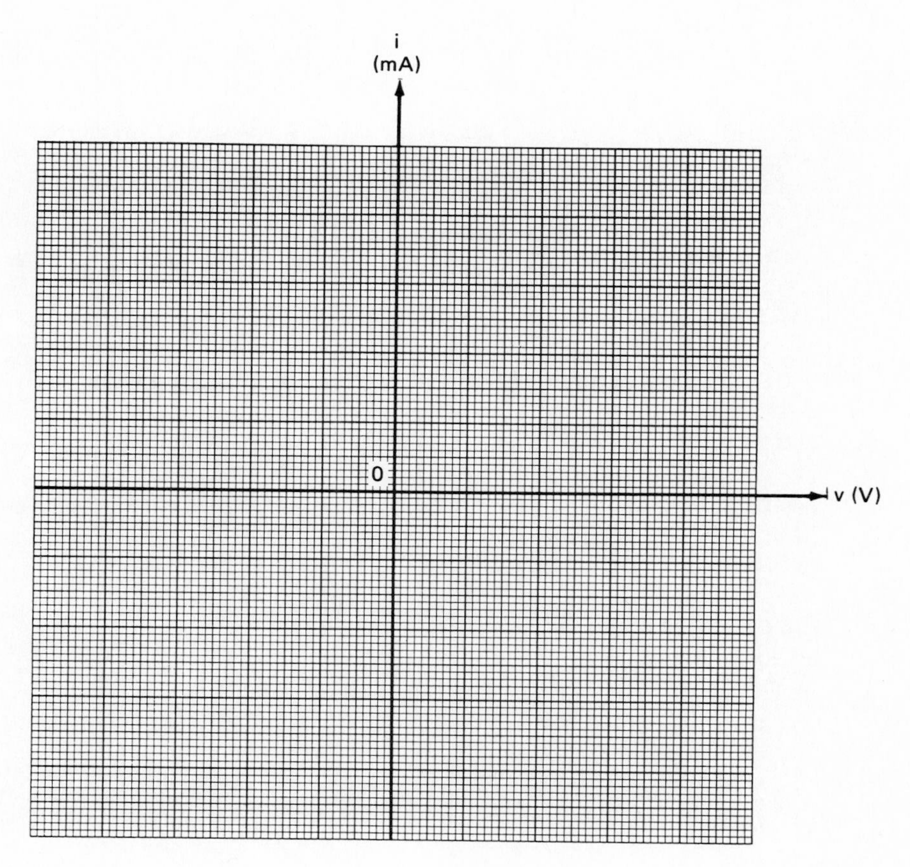

Graph 12.1

© 1987 by Prentice-Hall, Inc., A Division of Simon & Schuster, Englewood Cliffs, N.J. 07632. All rights reserved. Printed in the United States of America.

12.5 Comments and Conclusions:

1. What is the "inherent" error in the procedure for displaying the I-V characteristics?

2. Suppose that the source in Fig. 12.3 is a floating source, where would you connect the oscilloscope-ground terminal in order to display more accurately the I-V characteristics of any two-terminal device? Sketch the resulting display for a 1 kΩ resistor?

3. How is the I-V display of the semiconductor diode, Graph 12.1, affected by doubling the value of the current-limited resistance? Sketch the resulting display.

4. How is the I-V display of the 1 k Ω resistor, Graph 12.1, affected by doubling the (peak-to-peak) value of V_T? Sketch the resulting display.

© 1987 by Prentice-Hall, Inc., A Division of Simon & Schuster, Englewood Cliffs, N.J. 07632. All rights reserved. Printed in the United States of America.

Experiment 13

TRANSIENTS IN RC–CIRCUITS

Required Reading: Text, section 8.5

13.1 **Objective:**

To investigate the transient behaviour of simple RC-circuits.

13.2 **Prelab Assignment:**

Consider the circuit shown in Fig. 13.1. The switch S was in position (a) for a long time and then moved to position (b) at $t = o$. For each of the combinations of R & C shown in Table 13.1, determine:

(1) the circuit's time constant, τ ,

(2) the initial charging current, $i_C(o^+)$.

© 1987 by Prentice-Hall, Inc., A Division of Simon & Schuster, Englewood Cliffs, N.J. 07632. All rights reserved. Printed in the United States of America.

Table 13.1

Combination #	R [kΩ]	C [μF]
1	1	0.01
2	1	0.02
3	2	0.01

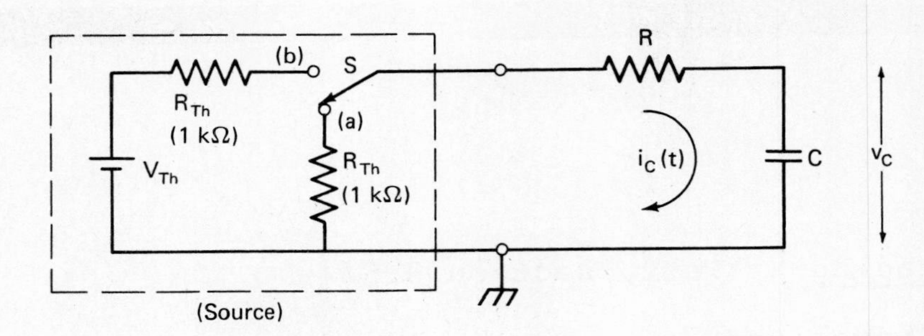

Fig. 13.1

Prelab Work Space:

Prelab Work Space:

13.3 Equipment:

ITEM	MANUFACTURER AND MODEL NO.	LAB. SERIAL NO
Dual-Beam Oscilloscope		
Decade Resistance Box		

Resistors: Two 1 kΩ

One 20 Ω

Capacitors: Two 0.01 μF

© 1987 by Prentice-Hall, Inc., A Division of Simon & Schuster, Englewood Cliffs, N.J. 07632. All rights reserved. Printed in the United States of America.

13.4 Procedure:

A. Initial Settings of the Oscilloscope:

Set the controls of the oscilloscope to the following positions:

. Trigger Source & mode	to	Y_A & Auto.
. (Time/div)	to	50 μ sec.
. (Y - position)	to	center of screen
. (DC - O - AC)	to	AC
. (V/div) for Y_A	to	0.1V
. (V/div) for Y_B	to	2 mV

B. Characterization of The Source (The CAL-output of The Oscilloscope):

(1) Connect (CAL-output) of the oscilloscope to Y_A. Adjust (Time/div) to display about one cycle of the source waveform, V_{Th}. Plot V_{Th} versus time on Graph 13.1.

(2) Measure the internal resistance, R_{Th} of the source by connecting the decade resistance box (R_x) across Y_A. Adjust R_x until the display is reduced to half of its initial value. Record V_{Th} and R_{Th} in Table 13.2.

4. What is the "inherent" error in the procedure for displaying v_C versus time?

5. Comment on your results; do your results verify the theoretical expectations? Explain the reasons for possible deviations.

© 1987 by Prentice-Hall, Inc., A Division of Simon & Schuster, Englewood Cliffs, N.J. 07632. All rights reserved. Printed in the United States of America.

Experiment 14

TRANSIENTS IN RL–CIRCUITS

Required Reading: Text, sections 10.3 & 10.4

14.1 **Objective:**

To investigate the transient behaviour of simple RL-circuits.

14.2 **Prelab Assignment:**

Consider the circuit shown in Fig. 14.1. The switch S was in position (a) for a long time and then moved to position (b) at $t = o$. For each of the combinations of R & L shown in Table 14.1, determine:

(1) the rise and decay time constants,

(2) the size of the current step through the inductor.

© 1987 by Prentice-Hall, Inc., A Division of Simon & Schuster, Englewood Cliffs, N.J. 07632. All rights reserved. Printed in the United States of America.

Table 14.1

Combination #	R [kΩ]	L [mH]
1	2	60
2	2	90
3	1	60

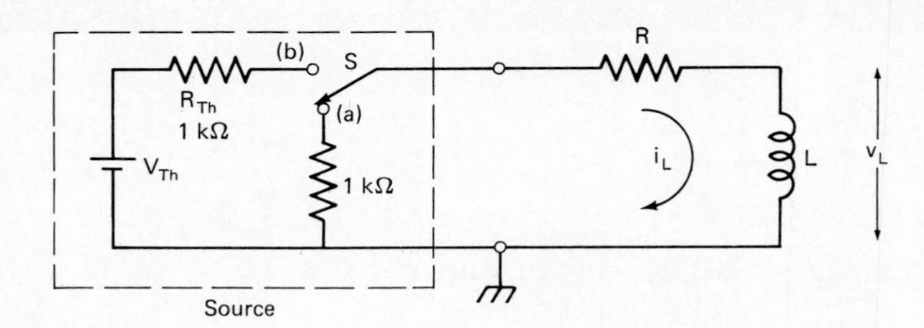

Fig. 14.1

Prelab Work Space:

Prelab Work Space:

14.3 Equipment:

ITEM	MANUFACTURER AND MODEL NO.	LAB. SERIAL NO
Dual-Beam Oscilloscope		
Decade Inductance Box		

Resistors: Two 1 kΩ

One 20 Ω

14.4 Procedure:

A. Initial Settings of the Oscilloscope:

Set the controls of the oscilloscope to the following positions:

© 1987 by Prentice-Hall, Inc., A Division of Simon & Schuster, Englewood Cliffs, N.J. 07632. All rights reserved. Printed in the United States of America.

- Trigger Source & mode to Y_A & Auto.
- (Time/div) to 50 μ sec.
- (DC - O - AC) to AC
- (V/div) for Y_A to 0.2V
- (V/div) for Y_B to 2 mV
- (Y - position) to center of screen

B. Transients Measurements:

(1) Connect the circuit shown in Fig. 14.2;

$R = 2\ k\Omega$ & $L = 60$ mH

[V_{Th} and R_{Th} represent the Thevenin's equivalent circuit of the (CAL) generator of the oscilloscope; See Section 13.4, Part B for the values of V_{Th} & R_{Th}.]

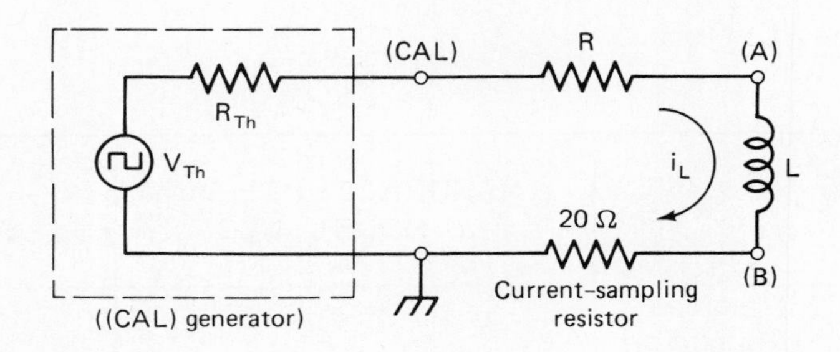

Fig. 14.2

(2) Connect node (A) to Y_A and node (B) to Y_B [the B-trace now displays 20x the current $i_L(t)$, while the A-trace displays (approximately) the voltage $v_L(t)$]. Plot v_L and i_L versus time on Graphs 14.1 and 14.2 respectively.

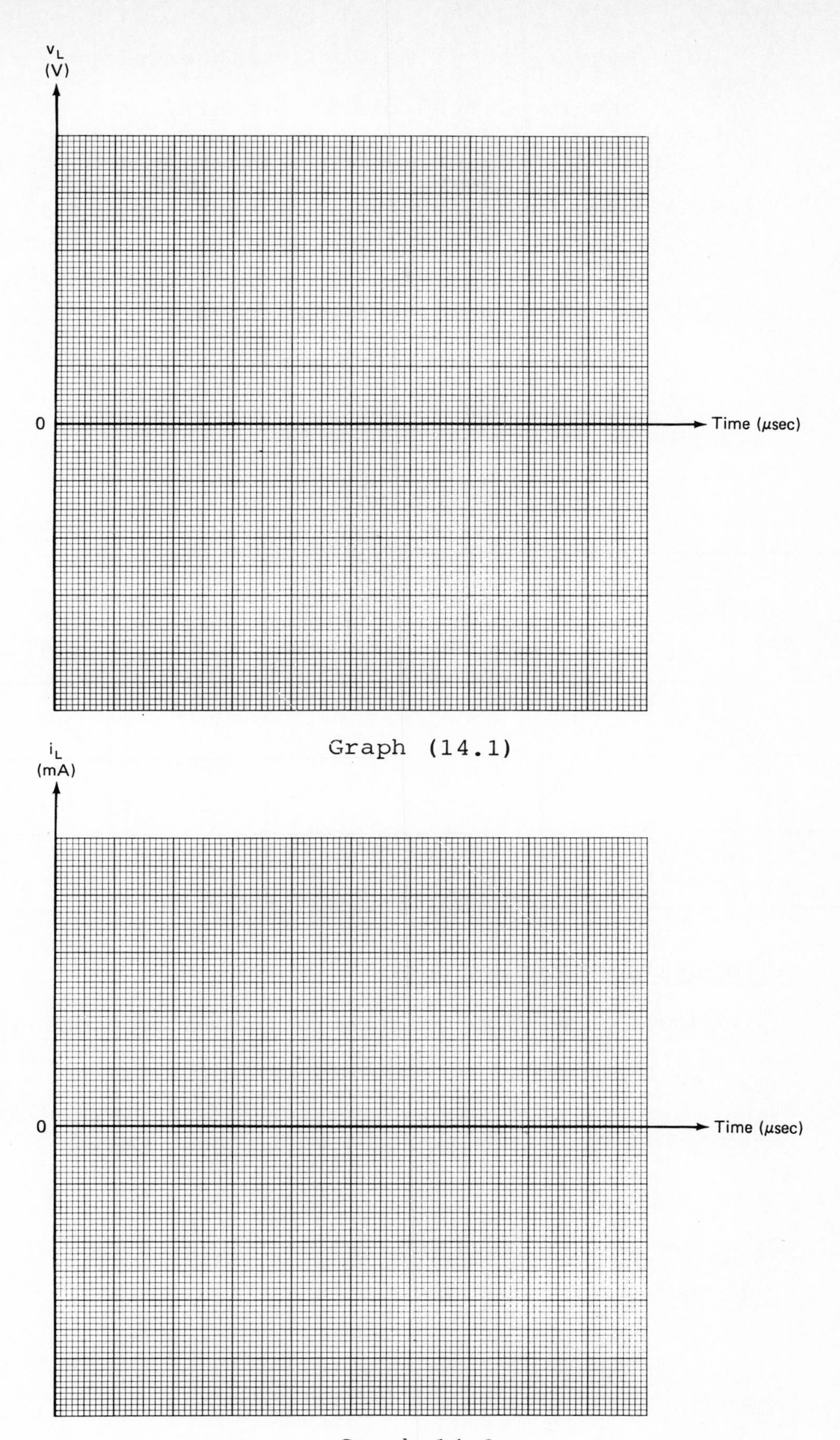

Graph (14.1)

Graph 14.2

© 1987 by Prentice-Hall, Inc., A Division of Simon & Schuster, Englewood Cliffs, N.J. 07632. All rights reserved. Printed in the United States of America.

(3) Adjust the (Time/div), trigger slope and level controls to measure, as accurately as you can, the rise and decay time constants. Record your results in Table 14.2.

(4) Repeat the above steps for all combinations of R & L given in Table 14.1.

Table 14.2

L (mH)	$R_T = R_{Th} + R$ [k Ω]	$\frac{L}{R_T}$ (μ sec)	τ_{rise} (μ sec)	τ_{dec} (μ sec)	Current Step [mA]

14.5 Comments and Conclusions:

1. How is the time constant affected by a change in L and R?

2. How is the final current i_L affected by a change in L and R?

3. How do the waveforms of $v_L(t)$ and $i_L(t)$ compare to those of $v_C(t)$ and $i_C(t)$ of Exp. #13? Explain.

4. Comment on your results; do your results verify the theoretical expectations? Explain the reasons for possible deviations.

© 1987 by Prentice-Hall, Inc., A Division of Simon & Schuster, Englewood Cliffs, N.J. 07632. All rights reserved. Printed in the United States of America.

Experiment 15

RESISTIVE AC–CIRCUITS

Required Reading: Text, sections 11.4 & 11.5

15.1 Objective:

- To examine the validity of Kirchhoff's laws for AC resistive circuits.
- To verify the RMS-peak-values relationship of sinusoidal waveforms.

15.2 Prelab Assignment:

Consider the circuit shown in Fig. 15.1.

$v_T(t)$ is a sinusoidal voltage waveform with a (p-p) value of 10V @ 200 Hz. Determine the following:

(1) the RMS value of the voltage of each node w.r.t. circuit ground,

(2) the RMS value of the current through each branch,

(3) the average power dissipated by each element.

© 1987 by Prentice-Hall, Inc., A Division of Simon & Schuster, Englewood Cliffs, N.J. 07632. All rights reserved. Printed in the United States of America.

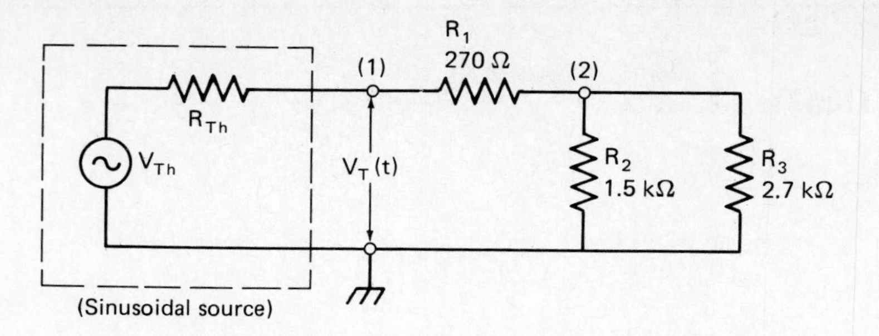

Fig. 15.1

Prelab Work Space:

15.3 Equipment:

ITEM	MANUFACTURER AND MODEL NO.	LAB. SERIAL NO
Dual-Beam Oscilloscope		
Signal generator		
DMM or VOM		
Decade Resistance Box		

Resistors: One 10 Ω, 270 Ω, 1.5k Ω & 2.7k Ω

15.4 Procedure:

A. Verification of Kirchhoff's law for AC Resistive Circuits:

(1) Connect the circuit shown in Fig. 15.2. Use channel Y_A of the oscilloscope to display the terminal voltage of the signal generator $v_T(t)$. Adjust V_T (p-p) to 10V @ 200 Hz.

(2) Use the DMM or the VOM to measure the RMS value of the voltage across the generator's terminals, R_1 and R_2. Estimate the RMS value of the current through each resistance and record your results in Table 15.1.

© 1987 by Prentice-Hall, Inc., A Division of Simon & Schuster, Englewood Cliffs, N.J. 07632. All rights reserved. Printed in the United States of America.

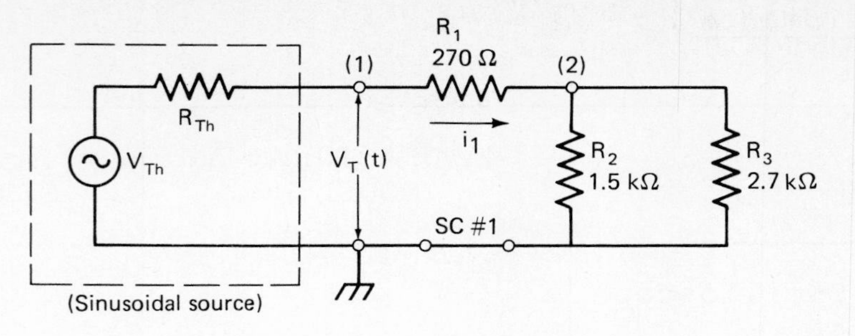

Fig. 15.2

Table 15.1

RMS - Values	Computed	Measured	Units
V_T			
V_1			
V_2			
I_1			
I_2			
I_3			

3. How does the measured value of V_T compare with the sum of the measured values of V_1 and V_2?

4. How does the estimated value of I_1 compare with the sum of the estimates of I_2 and I_3?

B. <u>Voltage and Current Waveforms for Resistive Elements:</u>

(1) Use Y_A to display $v_T(t)$; measure the peak value of $v_T(t)$.

V_T (peak) = ...

How does V_T (peak) compare with the measured RMS value of V_T?

(2) Replace SC # 1 with a $10\,\Omega$ resistance [this is the current-sampling resistance that provides a current-to-voltage conversion]. Use channel Y_B to display the voltage across the $10\,\Omega$ resistance. Y_B now displays 10 X the current waveform $i_1(t)$. Plot V_T and I_1 versus time on Graph 15.1.

© 1987 by Prentice-Hall, Inc., A Division of Simon & Schuster, Englewood Cliffs, N.J. 07632. All rights reserved. Printed in the United States of America.

(3) Estimate the value of $[v_T(t)]^2$ at eight equally-spaced time intervals within half a cycle. Plot your estimates of $[v_T(t)]^2$ versus time, over one cycle, on Graph 15.2.

Estimate, roughly, the total area under the curve representing $[v_T(t)]^2$, over one complete cycle.

Area = ...

(4) What is the average (mean) value of $[v_T(t)]^2$ over one complete cycle?

Mean $[v_T(t)]^2$ = ...

(5) What is the <u>root</u>-<u>mean</u>-<u>square</u> value of $v_T(t)$?

$\sqrt{\text{Mean } [v_T(t)]^2}$ = ...

How does the above result compare with the measure RMS value of V_T?

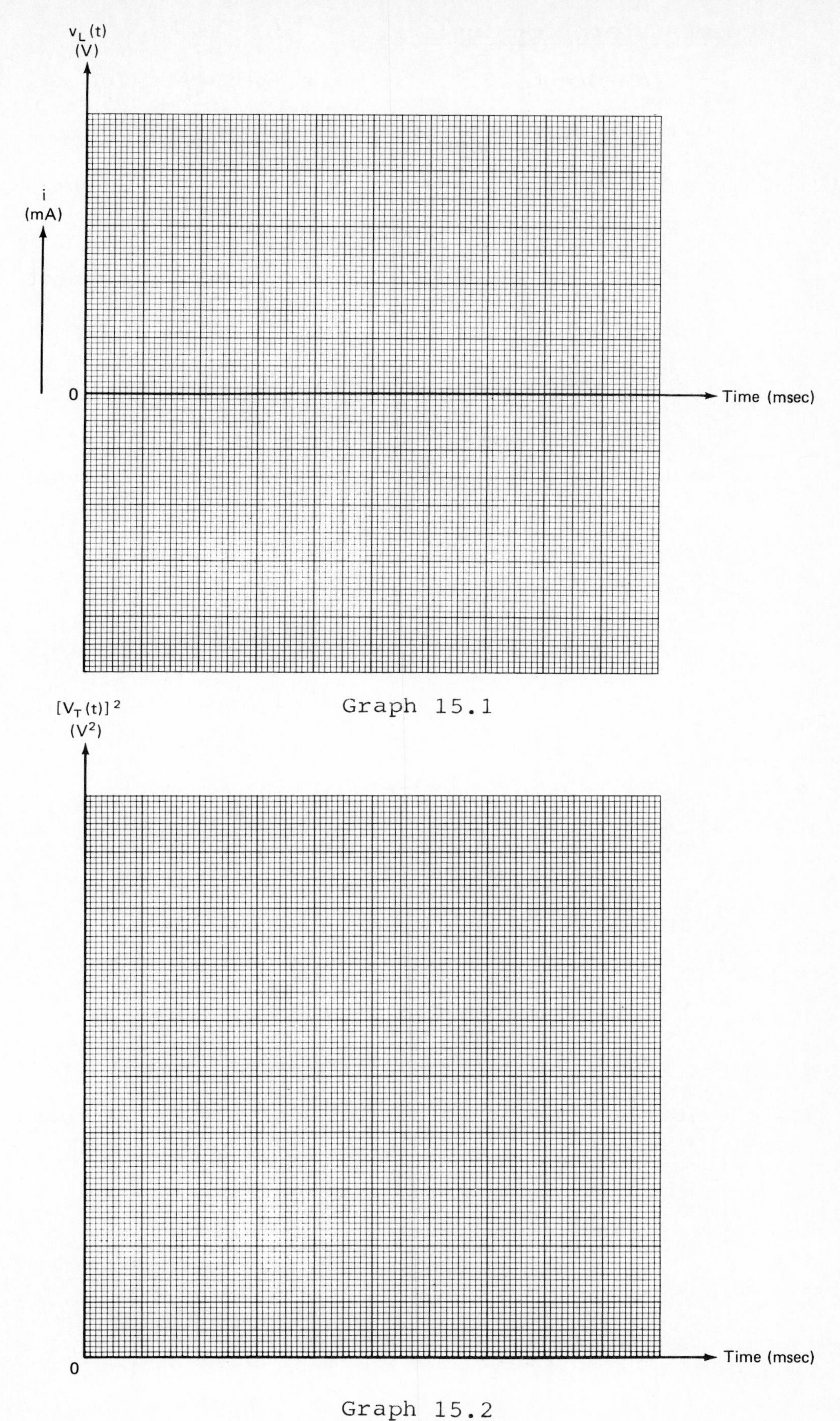

Graph 15.1

Graph 15.2

© 1987 by Prentice-Hall, Inc., A Division of Simon & Schuster, Englewood Cliffs, N.J. 07632. All rights reserved. Printed in the United States of America.

15.5 Comments and Conclusions:

1. From Graph 15.1, find the average value of $v_T(t)$ over one cycle.

   Mean $[v_T(t)]$ = ...

2. What is the theoretical ratio of the RMS-to-peak value of a sinusoidal waveform? What would this ratio be for an AC square waveform?

3. Suppose that the sinusoidal source in Fig. 15.2 is replaced by an AC triangular source, would Kirchhoff's laws still apply to your circuit?

4. Comment on your results; do the experimental results support the theoretical expectation? Explain the reasons for possible deviations.

Experiment 16

SERIES RL–CIRCUITS

<u>Required Reading:</u> Text, section 21.1

16.1 Objective:

- To examine the current-voltage relationship of an inductor for sinusoidal excitation at a constant frequency.
- To examine the current-voltage relationship of a series RL-circuit for sinusoidal excitation with varying frequency.

16.2 Prelab Assignment:

(1) Given a 0.5 H coil with unknown internal resistance r_L; its impedance triangle at 5 kHz has a phase angle of 84°. Determine r_L and the ratio of X_L to r_L at 5 kHz.

(2) A sinusoidal source and a 15 k Ω resistance are added in series with the above coil as shown in Fig. 16.1. The terminal voltage of

© 1987 by Prentice-Hall, Inc., A Division of Simon & Schuster, Englewood Cliffs, N.J. 07632. All rights reserved. Printed in the United States of America.

the source is maintained at 10V p-p, while the frequency is adjusted to each of the following values:

0.5, 1, 2, 5, 10 and 15 kHz.

a. Determine I (p-p) and the phase angle ($\phi°$) of V_T (w.r.t. I) for each of the above frequency values; calculate the magnitude of the input impedance $|Z_{in}|$ in each case.

b. Use Graphs 16.1 & 16.2 to plot $|Z_{in}|$ and ($\phi°$) versus frequency, respectively.

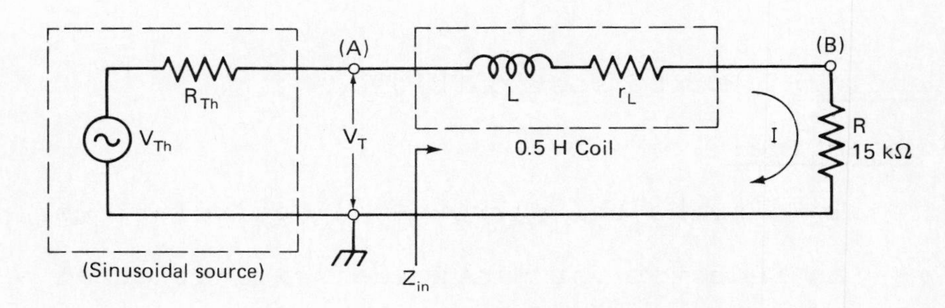

Fig. 16.1

Prelab Work Space:

Prelab Work Space:

© 1987 by Prentice-Hall, Inc., A Division of Simon & Schuster, Englewood Cliffs, N.J. 07632. All rights reserved. Printed in the United States of America.

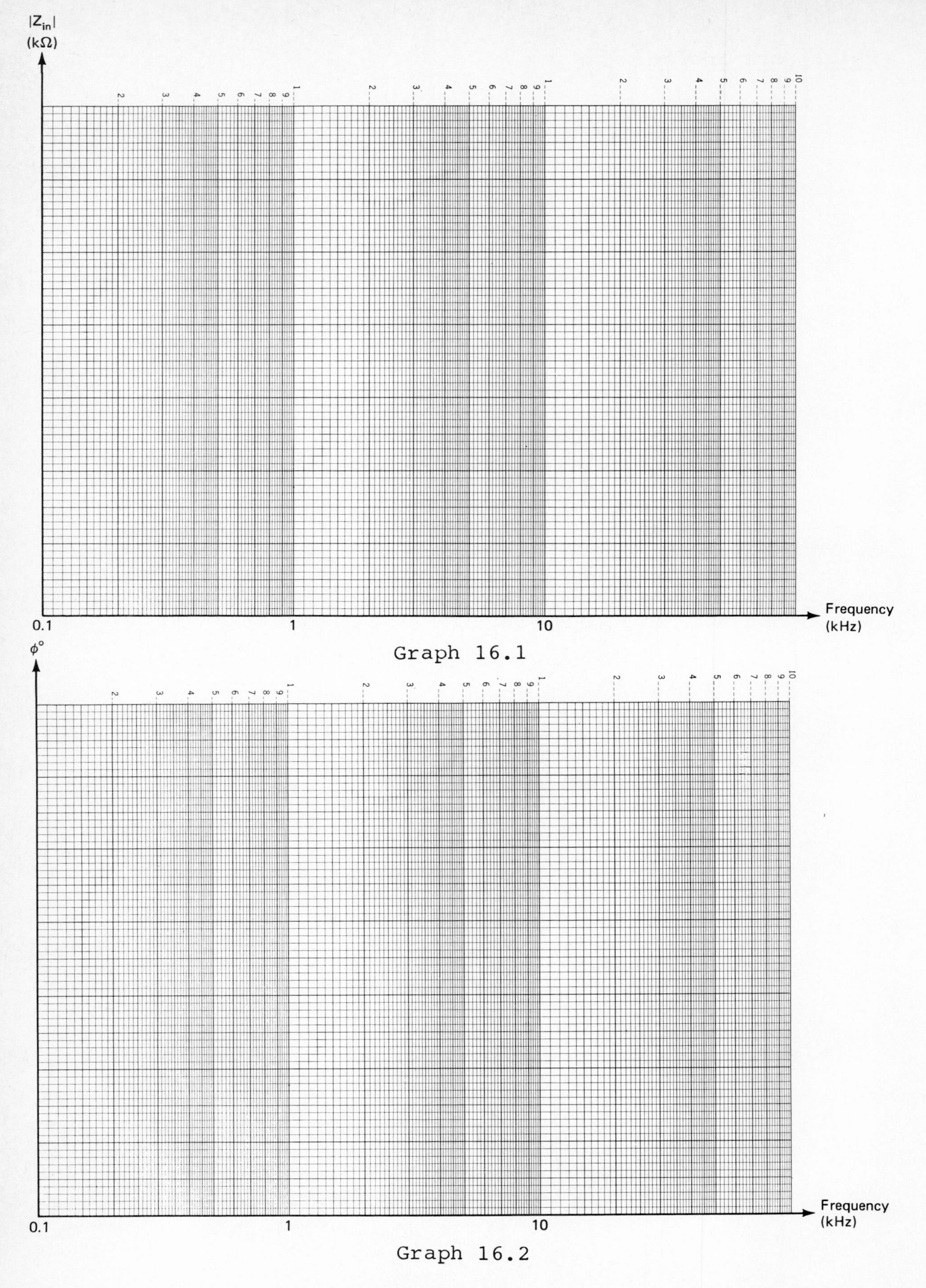

Graph 16.1

Graph 16.2

16.3 Equipment:

ITEM	MANUFACTURER AND MODEL NO.	LAB. SERIAL NO
Dual-Beam Oscilloscope		
Signal generator		
Decade Inductance Box		

Resistors: One 10 Ω & 15 k Ω

16.4 Procedure:

A. The I-V Relationship of an Inductor for Sinusoidal Excitation:

(1) Connect the circuit shown in Fig. 16.2. Use channel Y_A of the oscilloscope to display V_T; adjust V_T (p-p) to 10V at 5 kHz.

(2) Use channel Y_B to display the voltage across the 10 Ω resistor [V/div @ 2 mV]. Y_B now displays 10 X the current I, while Y_A displays (approximately) the voltage across the coil. Use Graph 16.3 to plot one cycle of both displays versus time.

© 1987 by Prentice-Hall, Inc., A Division of Simon & Schuster, Englewood Cliffs, N.J. 07632. All rights reserved. Printed in the United States of America.

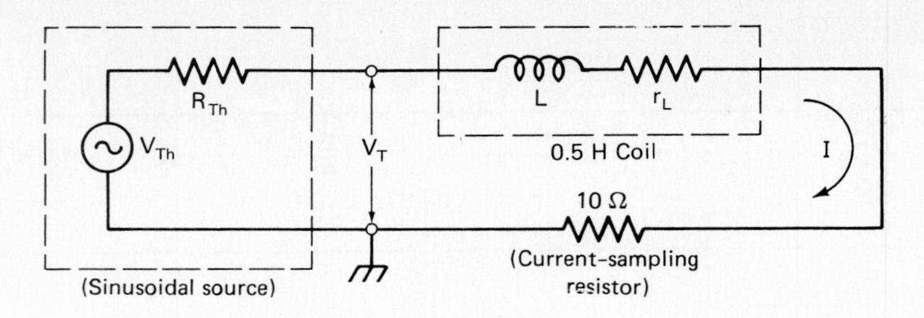

Fig. 16.2

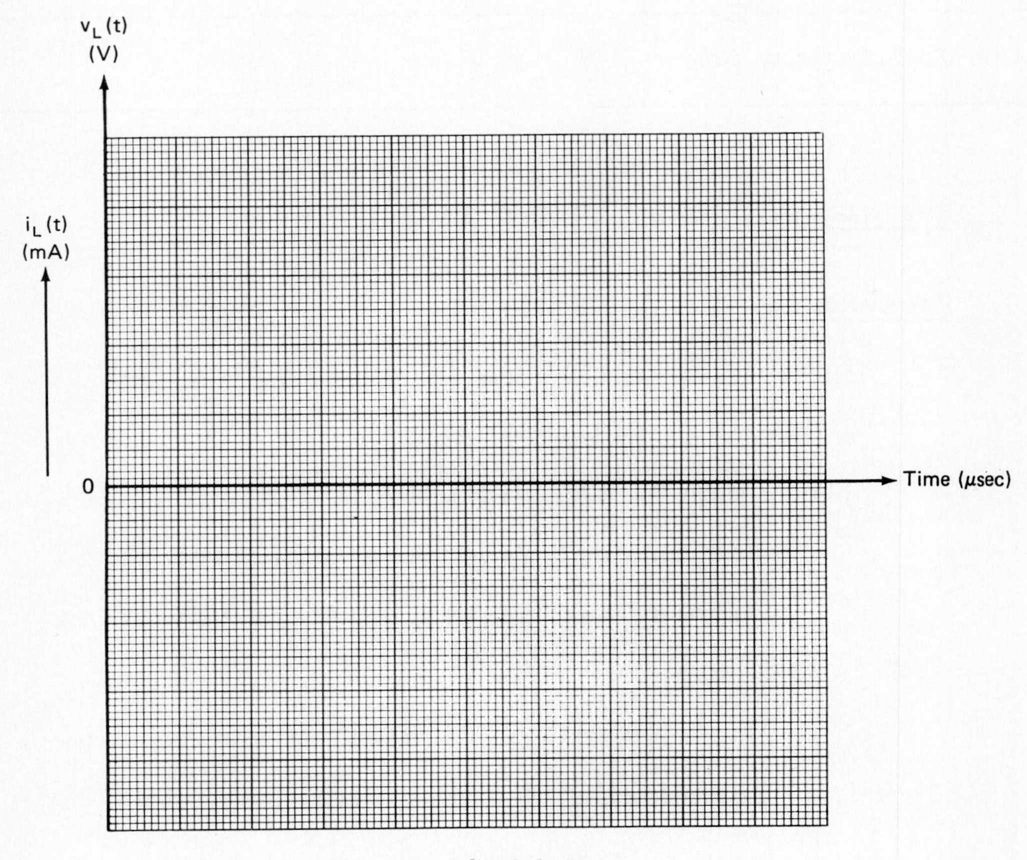

Graph 16.3

(3) Use Graph 16.4 to plot the phasor diagram (V & I) for the coil at 5 kHz; estimate the value of r_L and the ratio of X_L to r_L at 5 kHz.

r_L = ...

X_L/r_L = ...

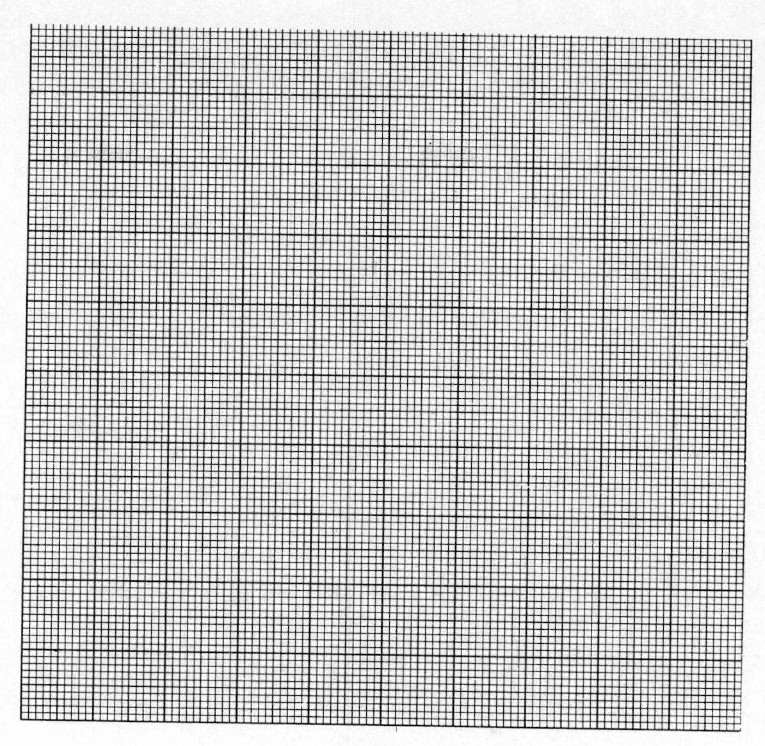

Graph 16.4

B. <u>Frequency Response of Simple RL-Circuit:</u>

(1) Connect the circuit shown in Fig. 16.1. Use Y_A to display V_T and Y_B to display the voltage of node (B) w.r.t. ground [Y_B now displays 15 k X I].

(2) Adjust V_T (p-p) to 10V at 5 kHz. Measure I (p-p) and the phase angle ($\phi°$) of V_T (w.r.t. I). [See Section 16.6, Appendix A: Phase Measurement.] Calculate the magnitude of the input impedance Z_{in}, and record in Table 16.1.

© 1987 by Prentice-Hall, Inc., A Division of Simon & Schuster, Englewood Cliffs, N.J. 07632. All rights reserved. Printed in the United States of America.

(3) Repeat step #2 for all the frequency settings shown in Table 16.1.

Table 16.1

Frequency (kHz)	0.5	1	2	5	10	15
I(p-p) (mA)						
$\|Z_{in}\|$ (k Ω)						
($\phi°$) (degrees)						

(3) Plot $|Z_{in}|$ and ($\phi°$) versus frequency on Graphs 16.1 & 16.2, respectively.

16.5 Comments and Conclusions:

1. How is the phase angle ($\phi°$) affected by:

   (a) a change in frequency,

   (b) a change in inductance,

   or (c) a change in resistance?

2. How is the magnitude of the input impedance $|Z_{in}|$ affected by:

(a) a change in frequency,

(b) a change in inductance,

or (c) a change in resistance?

3. Comment on your results; do your results agree with the theoretical expectations? Explain the reasons for possible deviations.

16.6 Appendix A: Phase Measurement

The following procedure is recommended for measuring the phase angle between two sinusoidal signals displayed by a dual-beam oscilloscope.

1. Set the controls of the oscilloscope to the following positions:

   .Trigger Source, mode & slope to Y_A, Auto & $(^+)$.

   .(Y - position) to center of screen

   .(DC - O - AC) to (AC)

2. Apply one sinusoid to Y_A and the other sinusoid to Y_B. Adjust both (V/div) such that

© 1987 by Prentice-Hall, Inc., A Division of Simon & Schuster, Englewood Cliffs, N.J. 07632. All rights reserved. Printed in the United States of America.

the (p-p) values of both sinusoids extend over the full height of the screen.

3. Adjust (Time/div) & (trigger slope and level) to set the display of one-half cycle of the Y_A-sinusoid to precisely nine horizontal divisions. [Each division now represents 20 degrees.]

4. Measure the phase angle ($\phi°$) of the Y_A-display w.r.t. the Y_B-display as:

   $\phi°$ [leading] = [No. of horizontal divisions between the positive zero-crossings of both displays] X 20°, as shown in Fig. 16.3a.

   $\phi°$ [lagging] = [No. of horizontal divisions between the negative zero-crossings of both displays] X 20°, as shown in Fig. 16.3b.

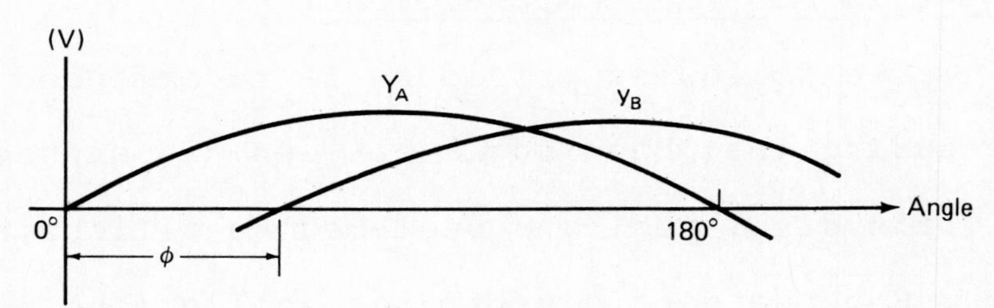

Fig. 16.3a

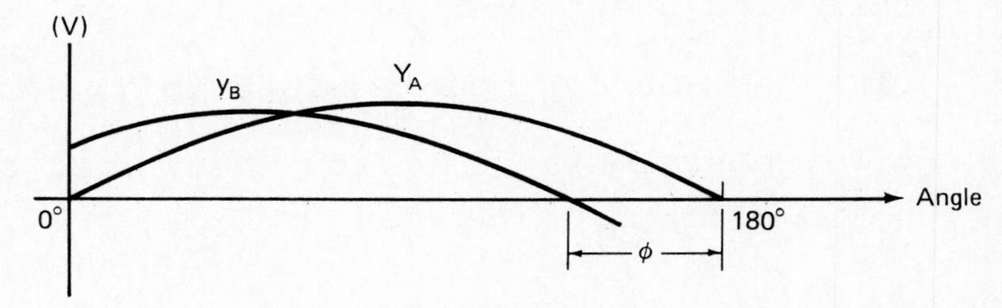

Fig. 16.3b

Experiment 17

SERIES RC–CIRCUITS

Required Reading: Text, section 12.2

17.1 Objective:

- To examine the current-voltage relationship of a capacitor for sinusoidal excitation at a constant frequency.
- To examine the current-voltage relationship of a series RC-circuit for sinusoidal excitation with varying frequency.

17.2 Prelab Assignment:

Consider the circuit shown in Fig. 17.1. The terminal voltage of the source is maintained at 10V p-p, while the frequency is adjusted to each of the following values:

0.5, 1, 2, 5, 10 and 15 kHz.

a. Determine I (p-p) and the phase angle ($\phi°$) of V_T (w.r.t. I) for each of the above

© 1987 by Prentice-Hall, Inc., A Division of Simon & Schuster, Englewood Cliffs, N.J. 07632. All rights reserved. Printed in the United States of America.

frequency values; calculate the magnitude of the input impedance $|Z_{in}|$ in each case.

b. Use Graphs 17.1 & 17.2 to plot $|Z_{in}|$ and ($\phi°$) versus frequency, respectively.

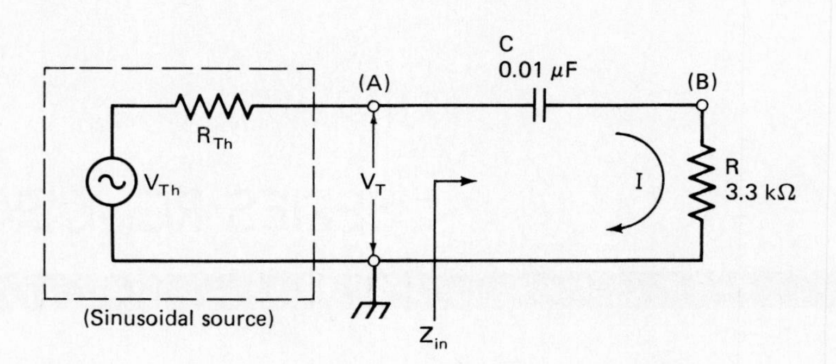

Fig. 17.1

Prelab Work Space:

Prelab Work Space:

© 1987 by Prentice-Hall, Inc., A Division of Simon & Schuster, Englewood Cliffs, N.J. 07632. All rights reserved. Printed in the United States of America.

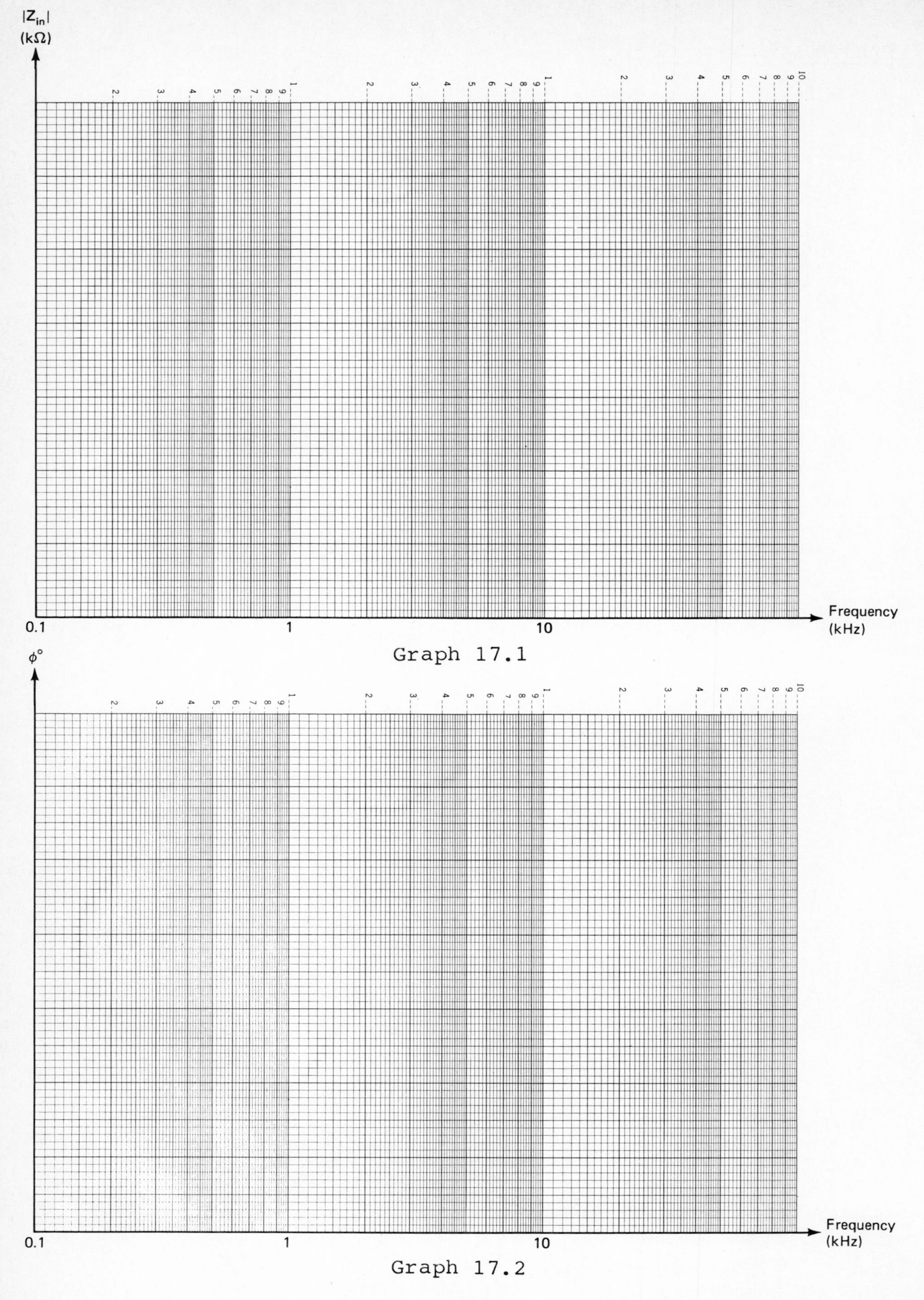

Graph 17.1

Graph 17.2

17.3 Equipment:

ITEM	MANUFACTURER AND MODEL NO.	LAB. SERIAL NO
Dual-Beam Oscilloscope		
Signal Generator		

Capacitors: One 0.01 μF

Resistors: One 10 Ω & 3.3 kΩ

17.4 Procedure:

A. The I-V Relationship of a Capacitor for Sinusoidal Excitation:

(1) Connect the circuit shown in Fig. 17.2. Use channel Y_A of the oscilloscope to display V_T; adjust V_T (p-p) to 10V at 5 kHz.

(2) Use channel Y_B to display the voltage across the 10 Ω resistor [V/div @ 5 mV]. Y_B now displays 10 X the current I, while Y_A displays (approximately) the voltage across the capacitor. Use Graph 17.3 to plot one cycle of both displays versus time.

(3) Use Graph 17.4 to plot the phasor diagram (V & I) for the capacitor at 5 kHz.

© 1987 by Prentice-Hall, Inc., A Division of Simon & Schuster, Englewood Cliffs, N.J. 07632. All rights reserved. Printed in the United States of America.

μ

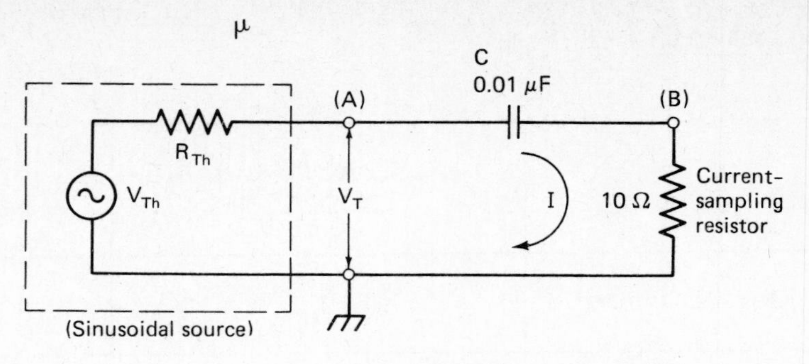

Fig. 17.2

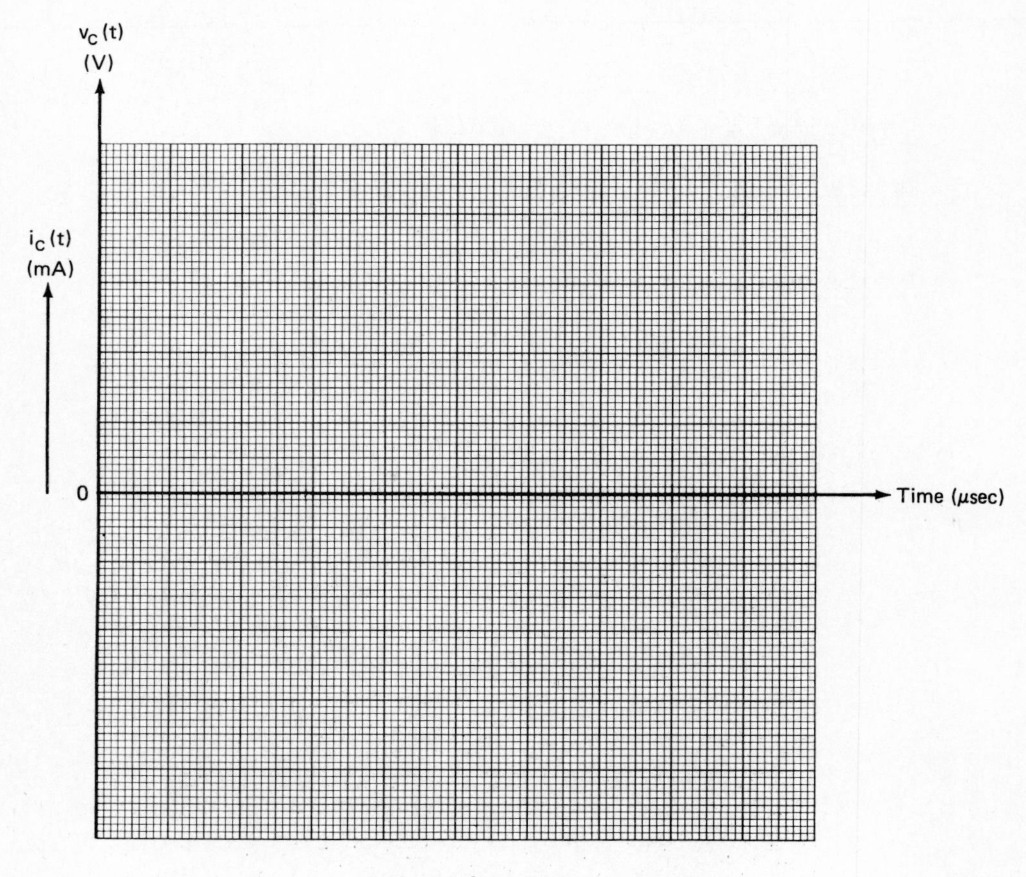

Graph 17.3

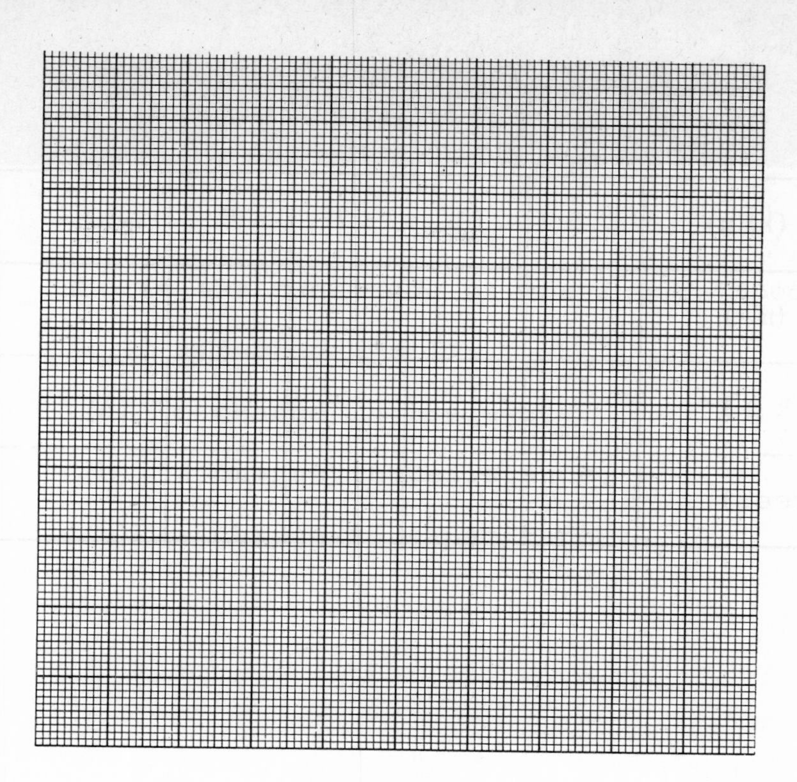

Graph 17.4

B. Frequency Response of Simple RC-Circuit:

(1) Connect the circuit shown in Fig. 17.1. Use Y_A to display V_T and Y_B to display the voltage of node (B) w.r.t. ground [Y_B now displays 3.3 k X I].

(2) Adjust V_T (p-p) to 10V at 5 kHz. Measure I (p-p) and the phase angle ($\phi°$) of V_T (w.r.t. I) [See Section 16.6.] Calculate the magnitude of the input impedance $|Z_{in}|$, and record in Table 17.1.

(3) Repeat step #2 for all the frequency settings shown in Table 17.1.

(4) Plot $|Z_{in}|$ and ($\phi°$) versus frequency on Graphs 17.1 & 17.2, respectively.

© 1987 by Prentice-Hall, Inc., A Division of Simon & Schuster, Englewood Cliffs, N.J. 07632. All rights reserved. Printed in the United States of America

© 1987 by Prentice-Hall, Inc., A Division of Simon & Schuster, Englewood Cliffs, N.J. 07632. All rights reserved. Printed in the United States of America.

Table 17.1

Experiment 18

SERIES R–L–C CIRCUITS

Required Reading: Text, section 12.3

18.1 Objective:

To examine the current-voltage relationship of a series R-L-C circuit for sinusoidal excitation at a varying frequency.

18.2 Prelab Assignment:

Consider the circuit shown in Fig. 18.1. The 0.5 H coil has an r_L of 1 kΩ at 2.25 kHz. The terminal voltage of the sinusoidal source is maintained at 10V (p-p), while the frequency is varied over the range: $0 < f < \infty$.

(1) Determine the frequency at which the magnitude of the current I is maximum.

(2) Determine the frequency at which the phase angle ($\phi°$) of V_T (w.r.t.I) is:

$0°$, $+45°$ and $-45°$

© 1987 by Prentice-Hall, Inc., A Division of Simon & Schuster, Englewood Cliffs, N.J. 07632. All rights reserved. Printed in the United States of America.

(3) Determine V_C (p-p) at resonance.

(4) Draw the phasor diagram [I, V_T, V_L and V_{r_L}] at: 2 kHz, 2.25 kHz and 2.5 kHz.

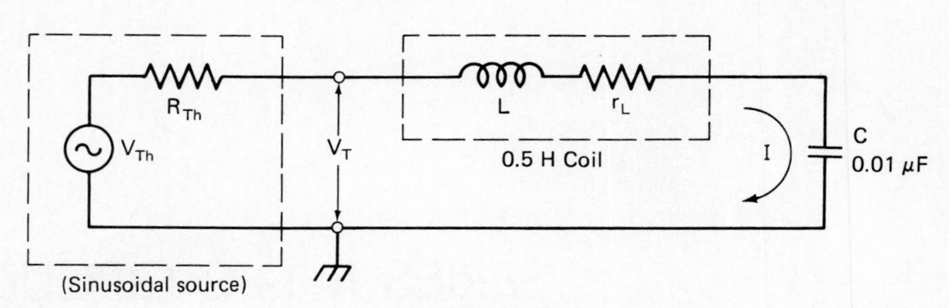

Fig. 18.1

Prelab Work Space:

Prelab Work Space:

© 1987 by Prentice-Hall, Inc., A Division of Simon & Schuster, Englewood Cliffs, N.J. 07632. All rights reserved. Printed in the United States of America.

18.3 Equipment:

ITEM	MANUFACTURER AND MODEL NO.	LAB. SERIAL NO
Dual-Beam Oscilloscope		
Signal generator		
Decade Inductance Box		

Resistors: One 10 Ω

Capacitors: Two 0.01 μF

18.4 Procedure:

(1) Connect the circuit shown in Fig. 18.2. Use channel Y_A of the oscilloscope to display V_T; adjust V_T (p-p) to 10V at any frequency (range 0.5 -- 10 kHz).

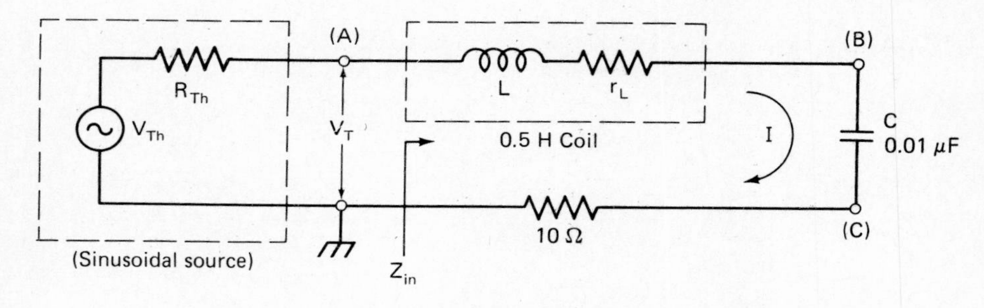

Fig. 18.2

(2) Use Y_B to display the voltage across the 10 Ω resistor. Y_B now displays 10 X I. Adjust the frequency of the signal generator until $|I|$ reaches its maximum value; this frequency is now referred to as "f_r". Measure I (p-p) and the phase angle ($\phi°$) of V_T w.r.t. I [See

Section 16.6]. Calculate the magnitude of the input impedance $|Z_{in}|$, and record your results in Table 18.1.

(3) Connect Y_B to display the voltage of node (B). Y_B now displays (approximately) the voltage across the capacitor V_C. Measure V_C (p-p). Calculate the magnitude of X_C and record in Table 18.1.

Table 18.1

Frequency	Hz	I (p-p) (mA)	$\|Z_{in}\|$ (kΩ)	$\phi°$ (°)	V_C (p-p) (V)	X_C (kΩ)
$0.2f_r$						
$0.5f_r$						
$0.8f_r$						
$0.9f_r$						
f_r						
$1.1f_r$						
$1.2f_r$						
$2f_r$						
$5f_r$						

(4) Connect Y_B again to display 10 X I and repeat the above set of measurements for all the

© 1987 by Prentice-Hall, Inc., A Division of Simon & Schuster, Englewood Cliffs, N.J. 07632. All rights reserved. Printed in the United States of America.

frequency settings shown in Table 18.1.

(5) Plot $|Z_{in}|$ and ($\phi°$) versus frequency on Graphs 18.1 and 18.2, respectively

(6) Plot the ratio of V_C(p-p)/V_T(p-p) versus frequency on Graph 18.3.

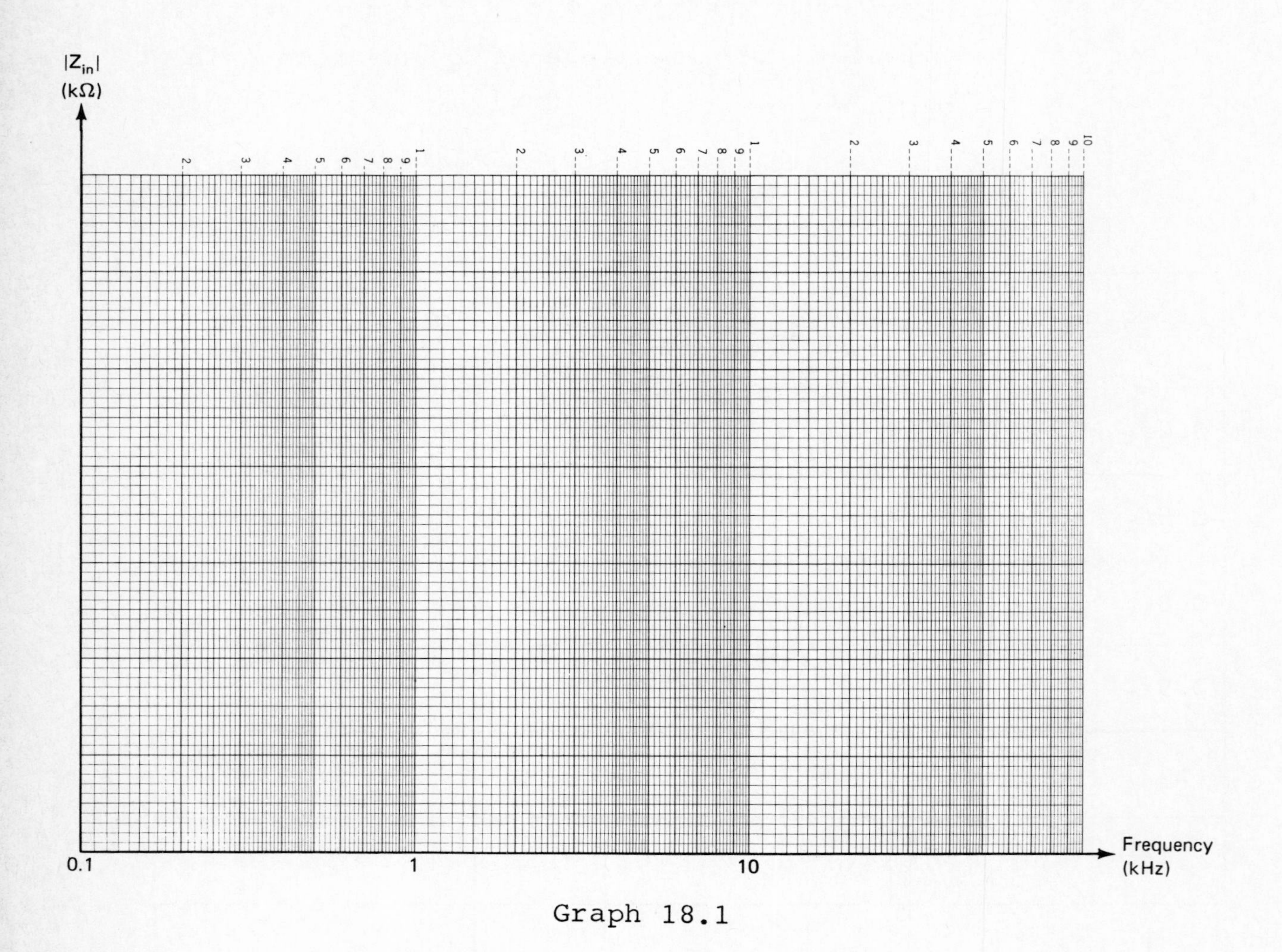

Graph 18.1

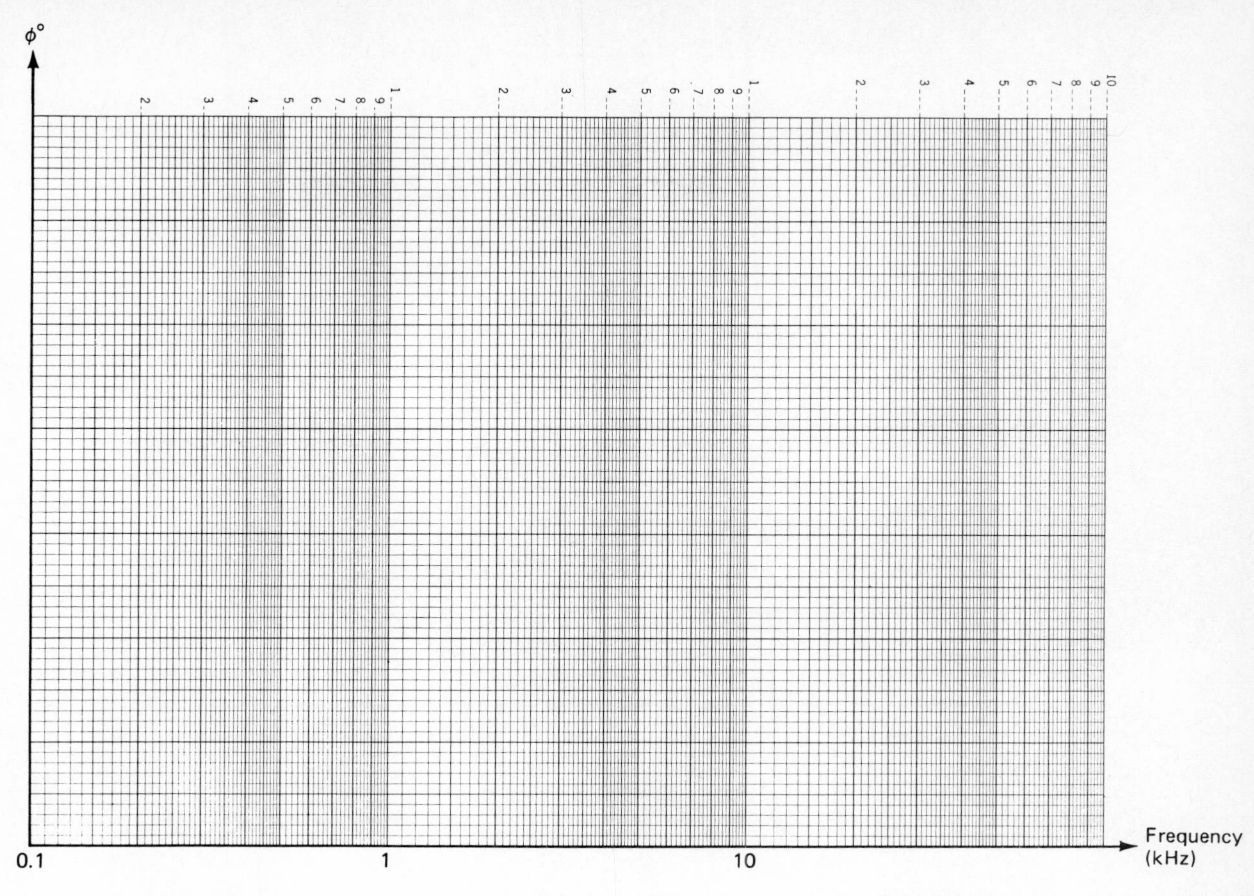

Graph 18.2

(7) Set the decade inductance box to 0.6 H. With Y_A displaying V_T and Y_B displaying 10 X I adjust the frequency of the signal generator until the circuit resonates; record f_r in Table 18.2.

(8) Repeat Step # 7 for:

(a) L = 0.4 H & C = 0.01 μ F

(b) L = 0.5 H & C = 0.02 μ F

(c) L = 0.5 H & C = 0.005 μ F

© 1987 by Prentice-Hall, Inc., A Division of Simon & Schuster, Englewood Cliffs, N.J. 07632. All rights reserved. Printed in the United States of America.

Table 18.2

L (H)	C (μF)	f_r (kHz)
0.5	0.01	
0.6	0.01	
0.4	0.01	
0.5	0.02	
0.5	0.005	

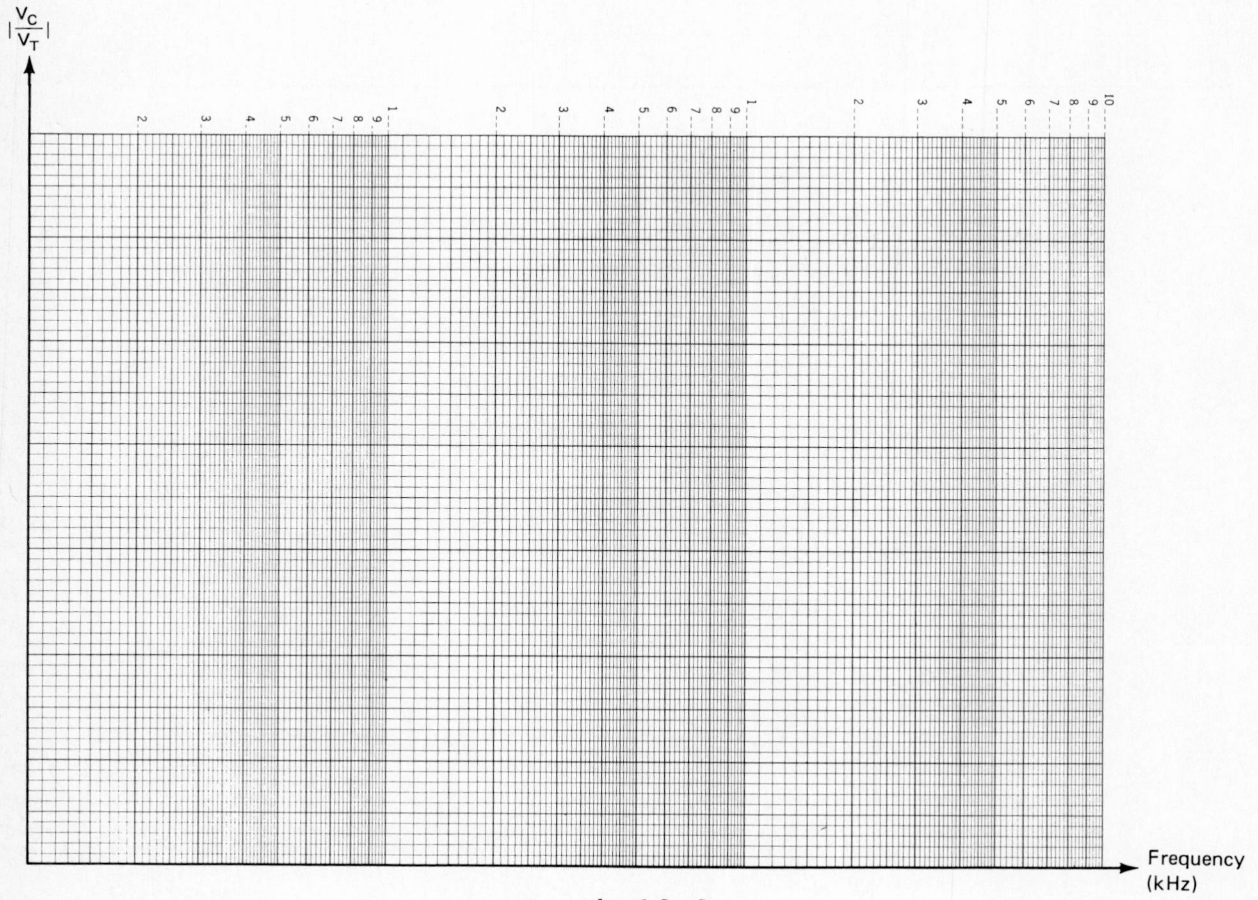

Graph 18.3

18.5 Comments and Conclusions:

1. Use Table 18.1 to determine the value of X_C, X_L and r_L at resonance.

2. How are the magnitude and phase of Z_{in} affected by the change in frequency?

3. Note that V_C at f_r is greater than the total applied voltage V_T. Does this mean that Kirchhoff's voltage law does not apply for series resonant circuits? Explain briefly.

© 1987 by Prentice-Hall, Inc., A Division of Simon & Schuster, Englewood Cliffs, N.J. 07632. All rights reserved. Printed in the United States of America.

4. How is f_r affected by a change in:

(a) L,

or (b) C?

5. Comment on your results; do your results support the theoretical expectations? Explain the reasons for possible deviations.

Experiment 19

PARALLEL R–L–C CIRCUITS

<u>Required Reading:</u> Text, section 12.4

19.1 **<u>Objective:</u>**

To examine the voltage-current relationship of a parallel R-L-C circuit for sinusoidal excitation with varying frequency.

19.2 **<u>Prelab Assignment:</u>**

A.. Consider the circuit shown in Fig. 19.1. The 5 mH coil has an internal resistance of r_L of 25 Ω at 2.25 kHz. Determine:

(1) the frequency at which Z_{in} is real (other than 0 Hz).

(2) the frequency at which $|Z_{in}|$ is maximum.

(3) the frequency at which $X_L = X_C$.

B. Repeat Part A when an additional resistance of 20 Ω is connected in series with the coil-section of the circuit.

© 1987 by Prentice-Hall, Inc., A Division of Simon & Schuster, Englewood Cliffs, N.J. 07632. All rights reserved. Printed in the United States of America.

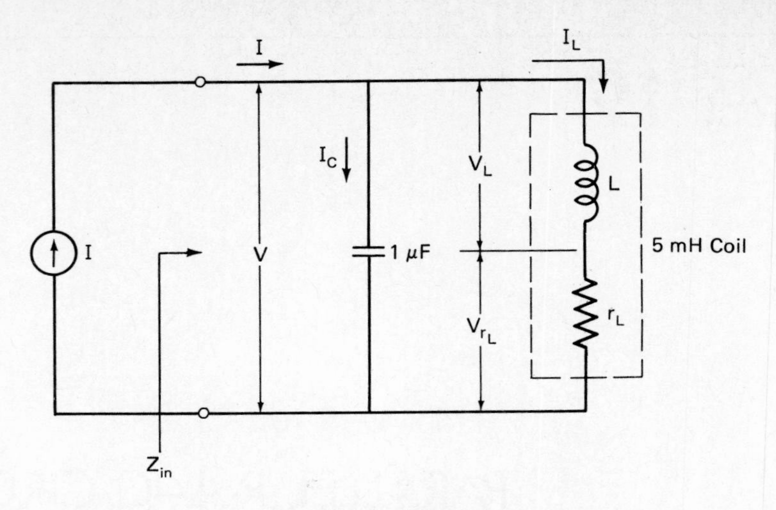

Fig. 19.1

C. Draw the phasor diagram for all the voltages and current in Fig. 19.1 at the frequency where Z_{in} is real.

Prelab Work Space:

Prelab Work Space:

© 1987 by Prentice-Hall, Inc., A Division of Simon & Schuster, Englewood Cliffs, N.J. 07632. All rights reserved. Printed in the United States of America.

19.3 **Equipment:**

ITEM	MANUFACTURER AND MODEL NO.	LAB. SERIAL NO
Dual-Beam Oscilloscope		
Signal Generator		
Decade Inductance Box		

Resistors: One 20 Ω & 10 kΩ

Capacitors: One 1 μF

19.4 **Procedure:**

(1) Connect the circuit shown in Fig. 19.2. Use channel Y_A of the oscilloscope to display V_T; adjust V_T (p-p) to 10V at any frequency (range 1 -- 10 kHz). [Note that since $|Z_{in}| \ll 10\ k\Omega$, I is approximately constant (independent of frequency).

$$I\ (p\text{-}p) = \frac{10}{(10k + Z_{in})} \approx \frac{10}{10k} = 1\ mA$$

[Thus, Y_A displays 10k x I]

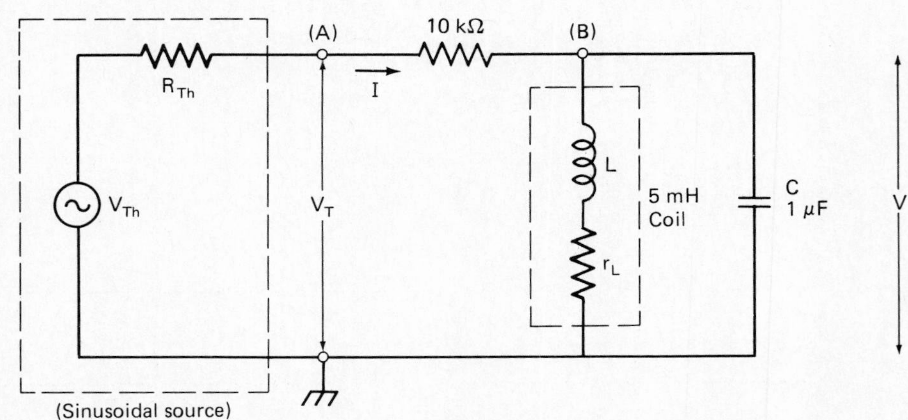

Fig. 19.2

(2) Connect Y_B to display the voltage of node (B). Y_B now displays the voltage V across the parallel tuned circuit. Adjust the frequency of the signal generator until I & V displays are in phase; this frequency is now referred to as f_o. Measure V (p-p) and determine $|Z_{in}|$.

(3) Adjust the frequency to each of the settings shown in Table 19.1; measure V (p-p) and the phase angle of V w.r.t. I. Calculate $|Z_{in}|$ and record your results.

Table 19.1

Frequency	(Hz)	V(p-p) (mV)	$\lvert Z_{in} \rvert$ (Ω)	$\phi°$ (°)
$0.1f_o$				
$0.5f_o$				
$0.8f_o$				
$0.9f_o$				
f_o				
$1.1f_o$				
$1.2f_o$				
$2f_o$				
$10f_o$				

© 1987 by Prentice-Hall, Inc., A Division of Simon & Schuster, Englewood Cliffs, N.J. 07632. All rights reserved. Printed in the United States of America.

(4) Adjust the frequency (on both sides of f_o) until Z_{in} drops to 0.707 Z_{in} max. The frequency difference between these two frequencies is referred to as the frequency bandwidth (BW) of the tuned circuit.

BW = ...

The ratio (f_o/BW) is a measure of the frequency selectivity of the tuned circuit; is known as the quality factor (Q).

Q = ...

(5) Plot Z_{in} and ($\phi°$) versus frequency on Graphs 19.1 & 19.2, respectively.

(6) Add a 20 Ω resistance in series with the coil-section of the circuit shown in Fig. 19.2. Repeat all the above measurements and record in Table 19.2.

BW = ...

Q = ...

Table 19.2

Frequency	(Hz)	V(p-p) (mV)	$\lvert Z_{in} \rvert$ (Ω)	$\phi°$ (°)
$0.1f_o$				
$0.5f_o$				
$0.8f_o$				
$0.9f_o$				
f_o				
$1.1f_o$				
$1.2f_o$				
$2f_o$				
$10f_o$				

(7) Plot Z_{in} and ($\phi°$) versus frequency on Graphs 19.1 & 19.2, respectively.

© 1987 by Prentice-Hall, Inc., A Division of Simon & Schuster, Englewood Cliffs, N.J. 07632. All rights reserved. Printed in the United States of America.

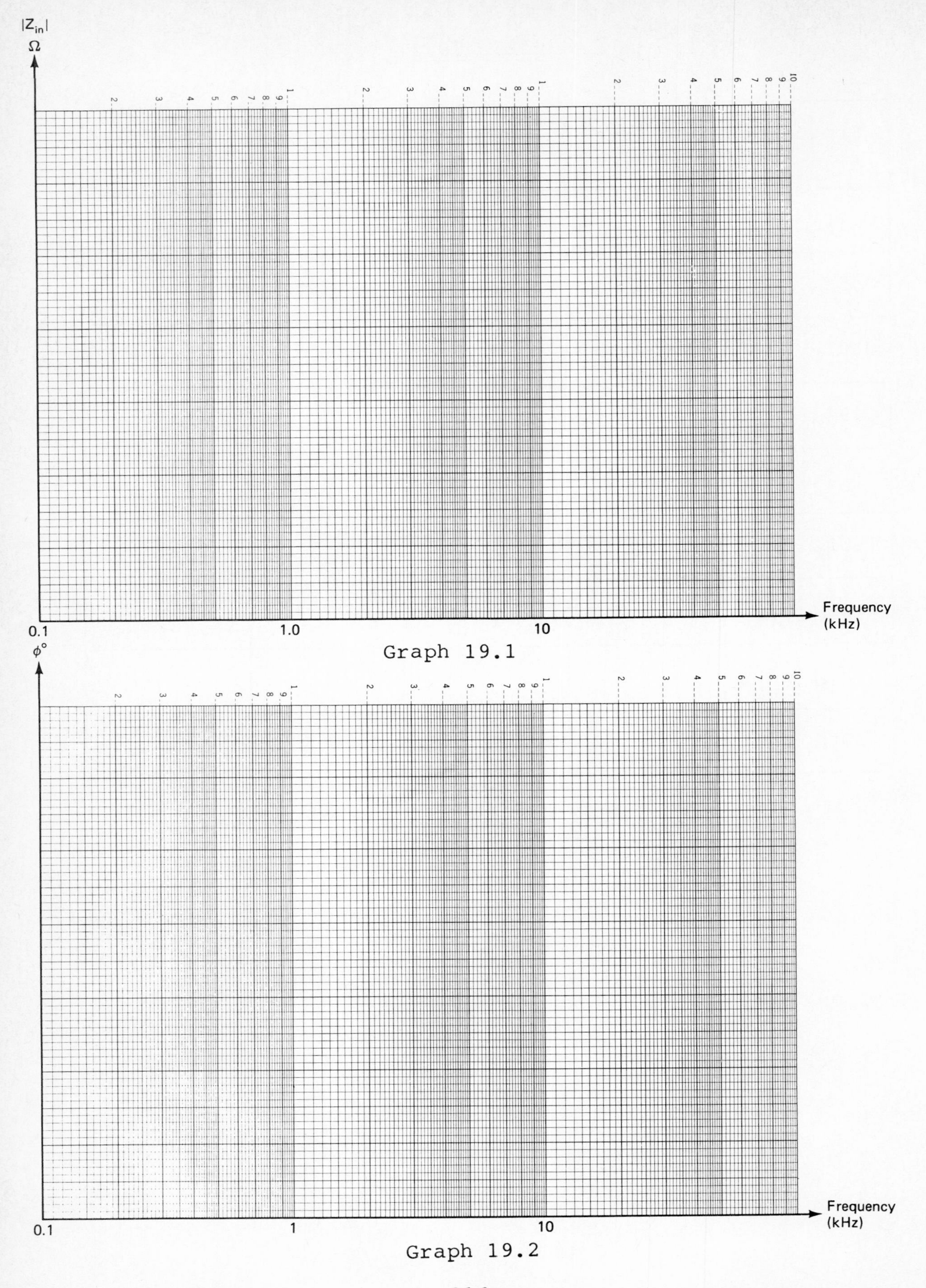

Graph 19.1

Graph 19.2

19.5 Comments and Conclusions:

1. How are the magnitude and phase of Z_{in} affected by the change in frequency?

2. How is f_o affected by:

   (a) a change in L or C,

   (b) adding extra resistance in series with the coil section?

3. How are the quality factor and bandwidth affected by the addition of the 20 Ω resistor?

© 1987 by Prentice-Hall, Inc., A Division of Simon & Schuster, Englewood Cliffs, N.J. 07632. All rights reserved. Printed in the United States of America.

4. Comment on your results; do your results support the theoretical expectations? Explain the reasons for possible deviations.

Experiment 20

SIMPLE FILTER CIRCUITS

Required Reading: Text, section 12.5

20.1 Objective:

To examine the frequency response (magnitude and phase) of simple low-pass and high-pass filters.

20.2 Prelab Assignment:

Consider the circuits shown in Fig. 20.1. The terminal voltage of the sinusoidal source is maintained at 1 V (p-p), while the frequency is varied.

For each circuit, determine the following:

a. the ratio of V_O/V_T as a function of ω (rad/sec).

© 1987 by Prentice-Hall, Inc., A Division of Simon & Schuster, Englewood Cliffs, N.J. 07632. All rights reserved. Printed in the United States of America.

b. the magnitude of V_O/ V_T at

$\omega = 0$ & $\omega = \infty$

c. the magnitude and phase of V_O/ V_T at

$\omega = 1/CR$ rad/sec.

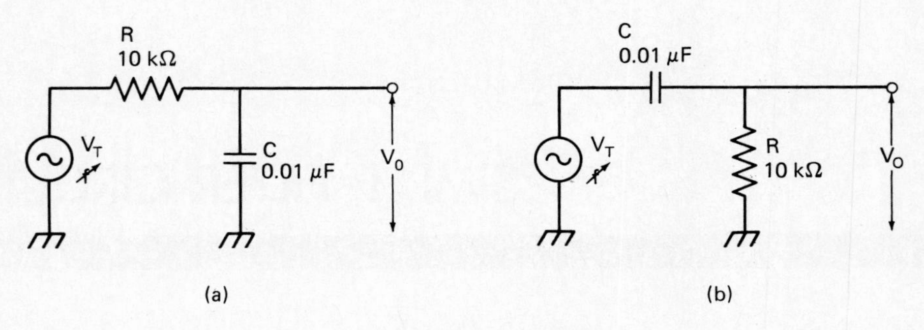

Fig. 20.1

Prelab Work Space:

Prelab Work Space:

© 1987 by Prentice-Hall, Inc., A Division of Simon & Schuster, Englewood Cliffs, N.J. 07632. All rights reserved. Printed in the United States of America.

20.3 **Equipment:**

ITEM	MANUFACTURER AND MODEL NO.	LAB. SERIAL NO
Dual-Beam Oscilloscope		
Signal generator		

Resistors: One 10 k Ω

Capacitors: One 0.01 μ F

20.4 **Procedure:**

A. The Frequency Response of A Simple Low-Pass Filter

(1) Connect the circuit shown in Fig. 20.2. Use channel Y_A of the oscilloscope to display V_T; adjust V_T (p-p) to 1 V at 100 Hz.

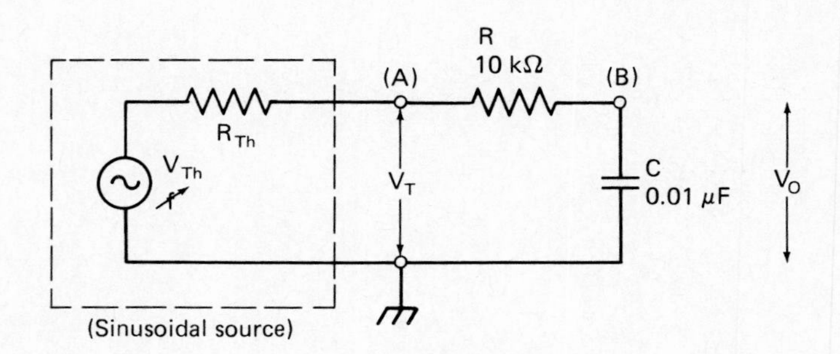

Fig. 20.2

(2) **Connect Y_B to display V_O. Measure V_O (p-p) and the phase angle ($\phi°$) of V_O w.r.t. V_T (See Section 16.6). Determine the ratio of V_O (p-p)/V_T (p-p) and record your results in Table 20.1**

(3) Repeat the above set of measurements for all the frequency settings shown in Table 20.1.

Table 20.1

Frequency (Hz)	100	500	1000	2000	5000	10,000
$\frac{V_O\ (p\text{-}p)}{V_T\ (p\text{-}p)}$ (V/V)						
$\phi°$ (degrees)						

(4) Plot V_O/V_T and ($\phi°$) versus frequency on Graphs 20.1 & 20.2, respectively. Determine the frequency (f_O) at which $V_O/V_T = 0.707$

B. <u>The Frequency Response of A Simple High-Pass Filter</u>

(1) Interchange R and C in the circuit shown in Fig. 20.2.

(2) Repeat as in Part (A) and record your results in Table 20.2.

© 1987 by Prentice-Hall, Inc., A Division of Simon & Schuster, Englewood Cliffs, N.J. 07632. All rights reserved. Printed in the United States of America.

Table 20.2

Frequency (Hz)	100	500	1000	2000	5000	10,000
$\frac{V_O\ (p-p)}{V_T\ (p-p)}$ (V/V)						
$\phi°$ (degrees)						

(3) Plot V_O/V_T and ($\phi°$) versus frequency on Graphs 20.1 & 20.2, respectively. Determine the frequency (f_O) at which $|V_O/V_T| = 0.707$

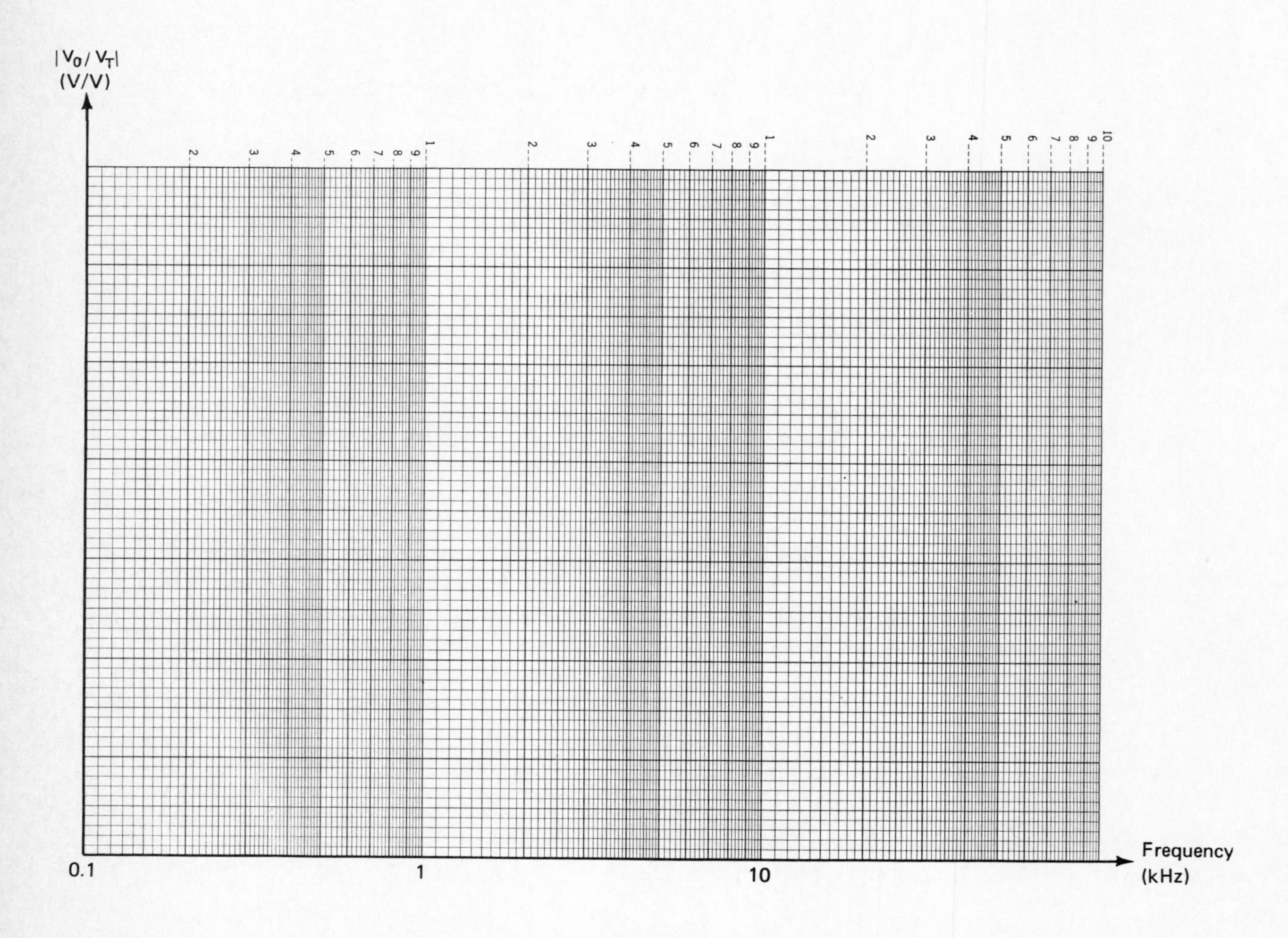

Graph 20.1

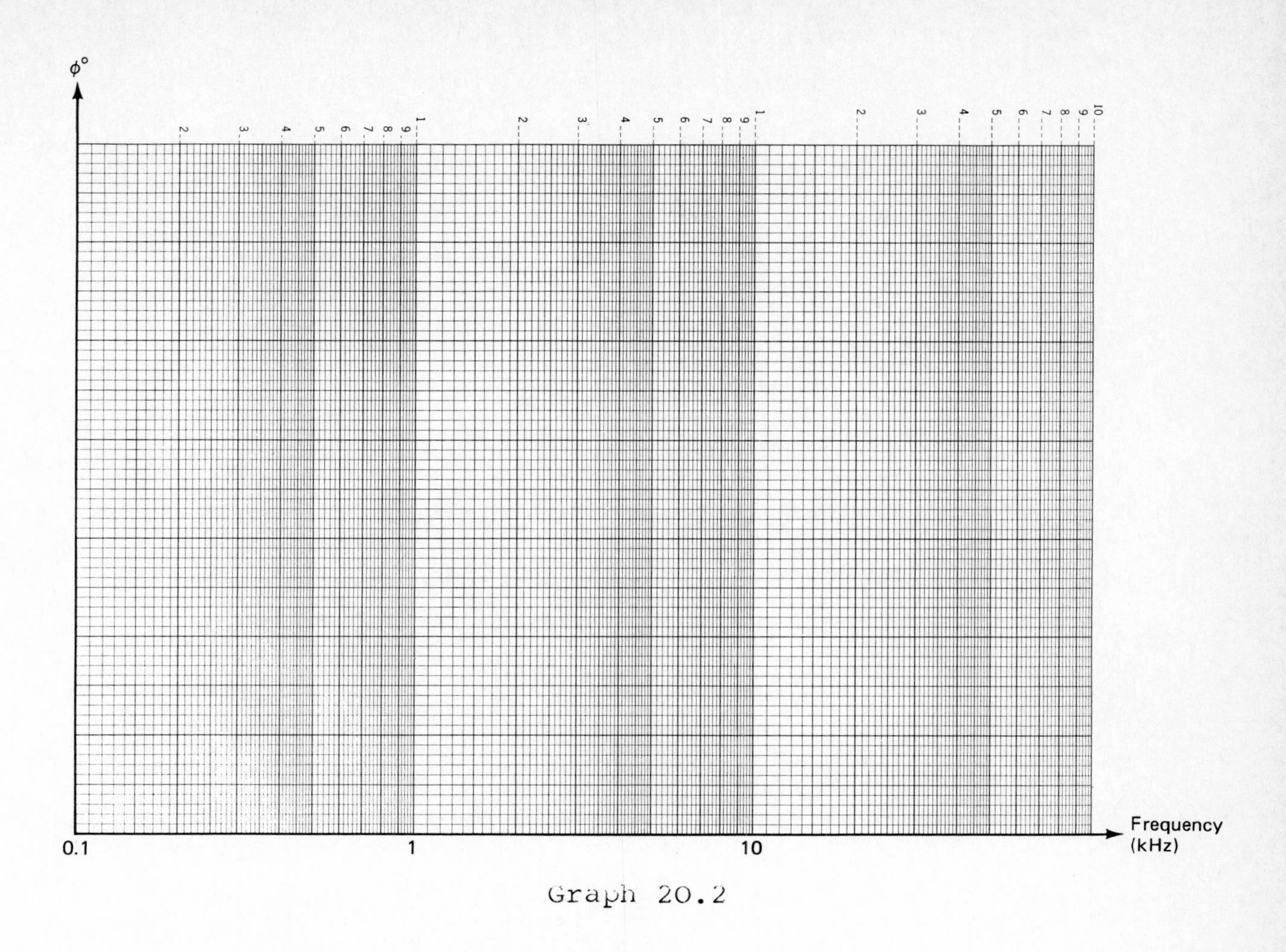

Graph 20.2

20.5 Comments and Conclusions:

A. Low-Pass Filter:

1. How are the magnitude and phase values of V_O/V_T affected by a change in frequency?

2. Suppose that (R) is replaced by (a R), and (C) is replaced by (C/a); a is a constant. How

© 1987 by Prentice-Hall, Inc., A Division of Simon & Schuster, Englewood Cliffs, N.J. 07632. All rights reserved. Printed in the United States of America.

would these changes affect the magnitude and phase of V_O/V_T ?

3. The Phase angle of V_O/V_T is negative at all frequencies. Explain qualitatively the reason for this behaviour.

B. <u>High-Pass Filter:</u>

1. How are the magnitude and phase values of V_O/V_T affected by a change in frequency?

2. The phase angle of V_O/V_T is positive at all frequencies. Explain (using a phasor diagram) the reason for this behaviour.

3. Comment on your results; do your results support the theoretical expectations? Explain the reasons for possible deviations.

© 1987 by Prentice-Hall, Inc., A Division of Simon & Schuster, Englewood Cliffs, N.J. 07632. All rights reserved. Printed in the United States of America.

Experiment 21

GENERAL AC–CIRCUIT: A LOW–PASS BUTTERWORTH FILTER

Required Reading: Text, section 12.5

21.1 Objective:

To examine the frequency response (magnitude and phase) of a simple series-parallel network.

21.2 Prelab Assignment:

Consider the network shown in Fig. 21.1. The network is a second-order low-pass Butterworth filter.

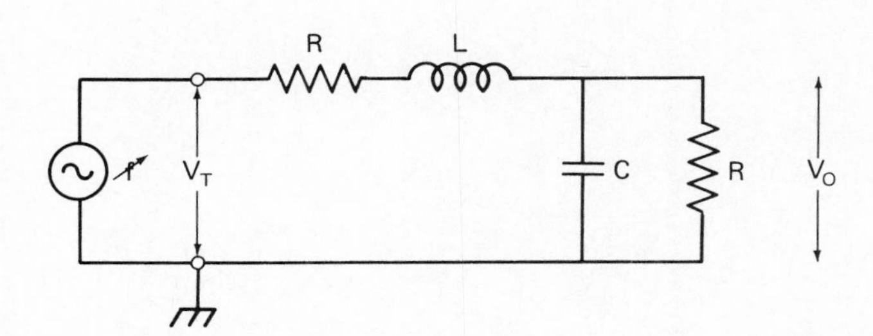

Fig. 21.1

© 1987 by Prentice-Hall, Inc., A Division of Simon & Schuster, Englewood Cliffs, N.J. 07632. All rights reserved. Printed in the United States of America.

(1) Determine the ratio of V_O/V_T as a function (rad/sec).

(2) Let R = 225 , L = 50 mH and C = 1 μF. Determine the magnitude and phase of V_O/V_T at each of the following frequency settings:

100, 500, 1000, 2000 & 5000 Hz.

(3) Plot the magnitude V_O/V_T and phase ($\phi°$) of V_O/V_T versus frequency on Graphs 21.1 & 21.2, respectively.

Prelab Work Space:

Prelab Work Space:

© 1987 by Prentice-Hall, Inc., A Division of Simon & Schuster, Englewood Cliffs, N.J. 07632. All rights reserved. Printed in the United States of America.

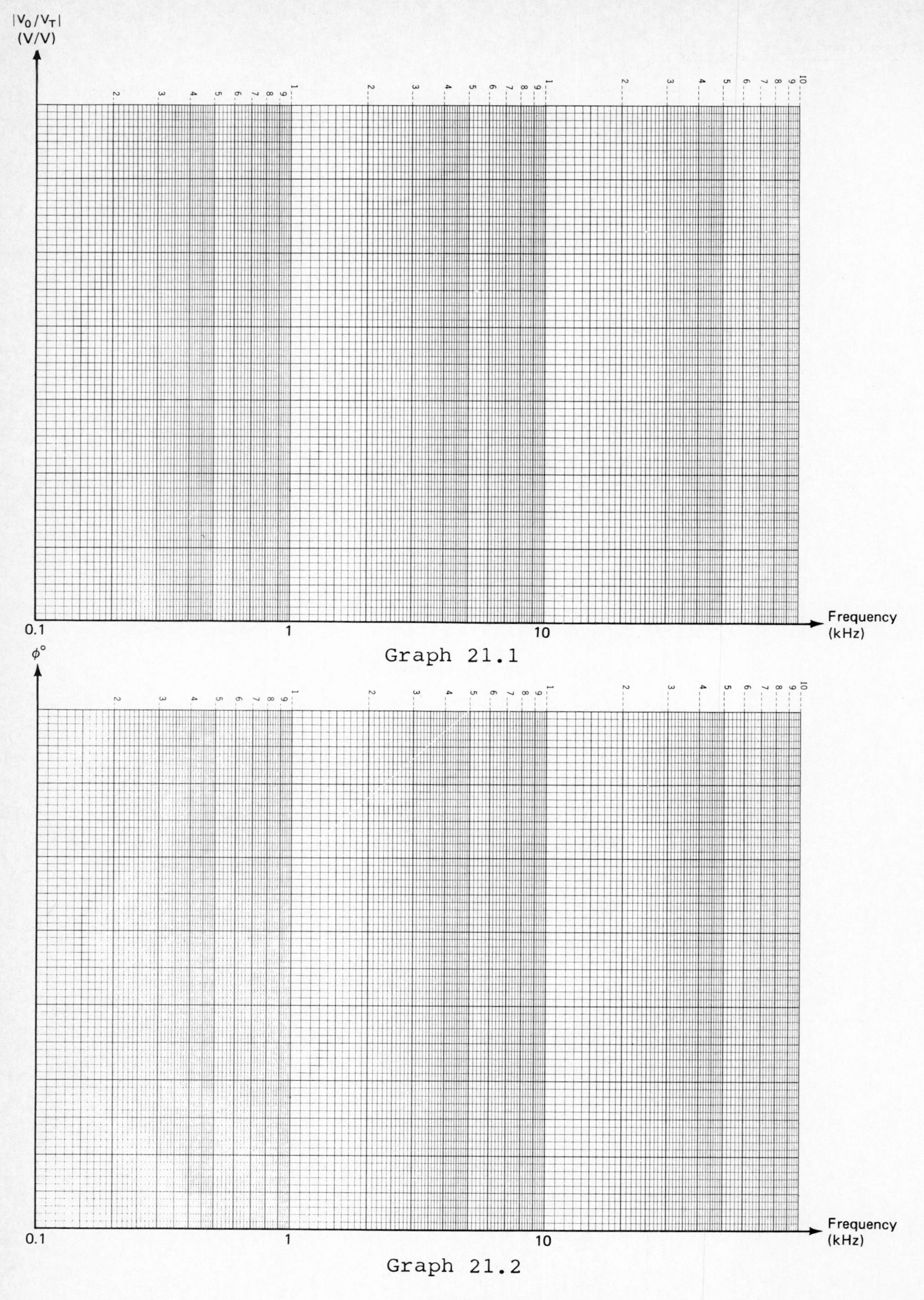

Graph 21.1

Graph 21.2

21.3 Equipment:

ITEM	MANUFACTURER AND MODEL NO.	LAB. SERIAL NO
Dual-Beam Oscilloscope		
Signal Generator		
VOM		
Two Decade Resistance Boxes		
Decade Inductance Box		

Capacitors: One 1 μ F

21.4 Procedure:

(1) Adjust the decade inductance box to 50 mH and use the VOM to measure its internal resistance r_L.

(2) Connect the circuit shown in Fig. 21.2. Adjust R_1 to 225 Ω and R_2 to $(225 - r_L)\,\Omega$.

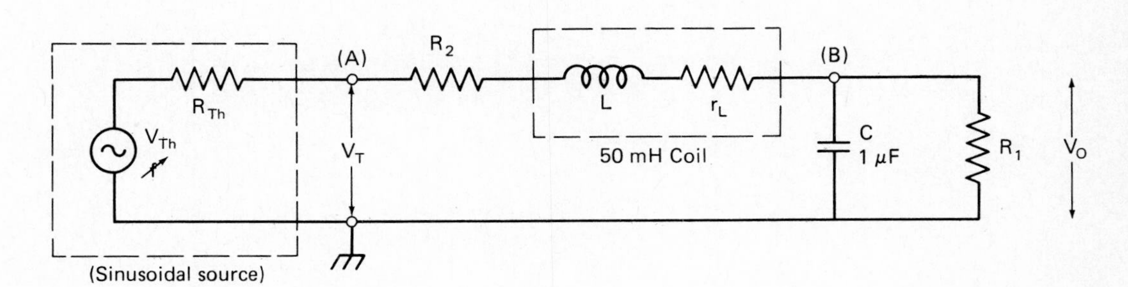

Fig. 21.2

© 1987 by Prentice-Hall, Inc., A Division of Simon & Schuster, Englewood Cliffs, N.J. 07632. All rights reserved. Printed in the United States of America.

(3) Use channel Y_A of the oscilloscope to display V_T; adjust V_T (p-p) to 10V at 100 Hz. [V_T must be maintained at 10V (p-p) for all the following frequency settings.]

(4) Use Y_B to display V_o. Measure V_o (p-p) and the phase angle ($\phi°$) of V_o w.r.t. V_T [See Section 16.6]. Determine the ratio of $\frac{V_o \text{ (p-p)}}{V_T \text{ (p-p)}}$, and record your results in Table 21.1.

(5) Repeat the above set of measurements for each frequency setting in Table 21.1.

Table 21.1

Frequency (Hz)	$\frac{V_O \text{ (p-p)}}{V_T \text{ (p-p)}}$ (V/V)	$\phi°$ (degrees)
100		
150		
250		
350		
500		
700		
1000		
2000		
3500		
5000		

(6) Plot V_O/V_T and ($\phi°$) versus frequency on Graphs 21.1 & 21.2, respectively.

© 1987 by Prentice-Hall, Inc., A Division of Simon & Schuster, Englewood Cliffs, N.J. 07632. All rights reserved. Printed in the United States of America.

21.5 Comments and Conclusions:

1. How are the magnitude and phase of V_O/V_T affected by a change in frequency?

2. The phase angle of V_O/V_T is negative at all frequencies. Explain qualitatively the reason for this behaviour.

3. Draw the phasor diagram for the voltage of each node (w.r.t. ground) and the current through each branch at 1000 Hz.

4. Repeat Question # 3 for f = 10 kHz. What are the major differences between the two phasor diagrams at 1 k and 10 kHz?

© 1987 by Prentice-Hall, Inc., A Division of Simon & Schuster, Englewood Cliffs, N.J. 07632. All rights reserved. Printed in the United States of America.

Experiment 22

THE TRANSFORMER

Required Reading: Text, section 17.5 and 17.6

22.1 **Objective:**

To understand the basic principles of transformers.

22.2 A. **Background:**

A transformer is a device that transfers electrical energy from one circuit to another by electromagnetic induction. Transformers need very little care and maintenance. They can vary in size from a very large stationary device such as a power transformer to miniature components using only a few windings and with air as the core material.

Transformers are more commonly used to step-up or step-down voltages. They are also frequently used to electrically isolate electronic equipment from the ac power source or isolate one part of the

© 1987 by Prentice-Hall, Inc., A Division of Simon & Schuster, Englewood Cliffs, N.J. 07632. All rights reserved. Printed in the United States of America.

circuit from the other; these transformers are called 'isolation transformers'. Transformers are also used in some electronic circuits to match the impedance of the load to the source impedance.

Transformers are primarily classified according to their usage such as audio, power, radio frequency, modulation and filament transformers. They can also be classified according to the type and shape of the core material used.

The typical transformer has two windings: a 'primary' which is connected to the source and a 'secondary' across which the load is connected. They are wound on a common magnetic core and insulated from each other electrically.

The following relationships are valid under ideal operating conditions of transformers:

$$\frac{V_p}{V_s} = \frac{I_s}{I_p} = \frac{N_p}{N_s} = a$$

where, 'a' is the transformation or turns ratio.

The impedance of the secondary circuit reflected to the primary side is determined from:

$$Z_p = a^2 Z_s$$

The equation above can be used to calculate the turns ratio required to match the reflected impedance of the load to the source impedance.

22.2 B. **Prelab Assignment:**

(1) A transformer rated at 50 volt-amps, 120V (Primary)/50V (Secondary), is connected to a 60 Hz AC source with a terminal voltage of 20 Volts RMS. Calculate:

(a) the primary and secondary currents when a load resistance of 50 Ω is connected across the secondary winding.

(b) the number of turns on the secondary winding if the primary winding has 720 turns.

(c) the RMS current in the primary winding if a resistive load of 500 Ω is connected across the secondary; determine the value of the resistance reflected to the primary side.

(d) the reflected impedance if a load of 10 μ F capacitor in series with 200 Ω resistance is connected across the secondary winding.

(2) Assume that the above transformer has a center-tapped secondary winding and a load of 50 Ω connected as shown in Fig. 22.1. Calculate:

(a) the RMS values of V_S, I_P and I_S,

(b) 'a' and Z_p.

(c) Volt-Ampere (input) and Volt-Ampere (Output).

Record your results in Table 22.2.

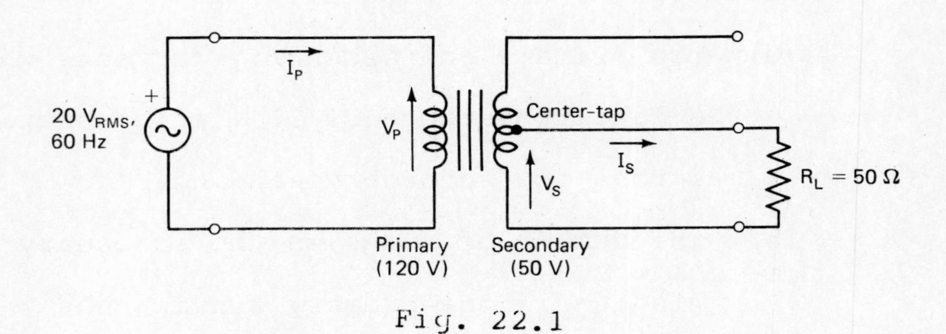

Fig. 22.1

Prelab Work Space:

Prelab Work Space:

© 1987 by Prentice-Hall, Inc., A Division of Simon & Schuster, Englewood Cliffs, N.J. 07632. All rights reserved. Printed in the United States of America.

22.3 Equipment:

ITEM	MANUFACTURER AND MODEL NO.	LAB. SERIAL NO
Signal Generator		
DMM or VOM		
AC Milliammeter		
Decade Resistance Box		
Transformer		

Capacitors: 10 μF

22.4 Procedure:

(1) Connect the circuit shown in Fig. 22.2. Adjust the terminal voltage of the AC source to 20V RMS at 60 Hz.

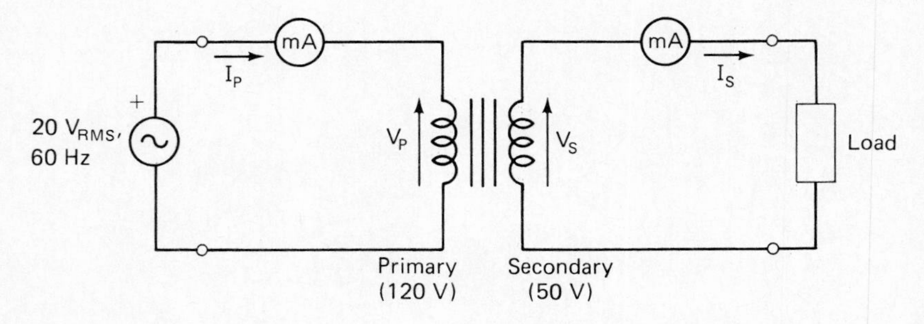

Fig. 22.2

(2) For the various values of the loads shown in Table 22.1, measure the RMS values of V_p, V_s, I_p and I_s. Record the results in Table 22.1.

Table 22.1

Load	V_p (V)	V_s (V)	I_p (mA)	I_s (mA)	$\frac{V_p}{V_s}$	$\frac{I_s}{I_p}$	Z_p (Ω)	Z_s (Ω)
Open-Circuit								
R_L = 1 kΩ								
R_L = 500 Ω								
R_L = 100 Ω								
R_L = 50 Ω								
10 μF in Series with 200 Ω								

(3) Connect the circuit shown in Fig. 22.3.

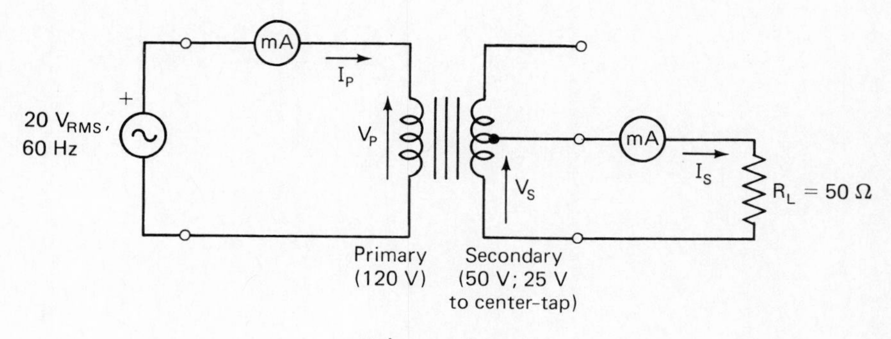

Fig. 22.3

© 1987 by Prentice-Hall, Inc., A Division of Simon & Schuster, Englewood Cliffs, N.J. 07632. All rights reserved. Printed in the United States of America.

Measure the RMS values of V_p, V_s, I_s and I_p. Record the results in Table 22.2 and calculate 'a', Z_p, Volt-Ampere (input) and Volt-Ampere (output).

Table 22.2

	V_p (V)	V_s (V)	I_p (mA)	I_s (mA)	$a=\frac{V_p}{V_s}$	Z_p (Ω)	(VA) IN	(VA) OUT
CALC. Prelab.								
Measured								

22.5 Comments and Conclusions:

1. Did the measured values of the transformation ratio [as determined by the voltage and current measurements in Table 22.2] match the name plate ratio of transformation? If not, what are the possible reasons?

2. Do the values of Z_p calculated (for $R_L=50\,\Omega$) in Tables 22.1 and 22.2 equal the corresponding values of a^2Z_s? If not, explain why.

3. What conclusions can you draw from the power calculations in Table 22.2?

4. If a transformer is required to be used as an isolation transformer only, what should be the ratio of transformation (a)?

© 1987 by Prentice-Hall, Inc., A Division of Simon & Schuster, Englewood Cliffs, N.J. 07632. All rights reserved. Printed in the United States of Amer

5. What is the 'rated' secondary current of your transformer if it is used as:

(a) a step-up transformer,

(b) a step-down transformer?